7× table	8× table	9× table	10× table	11× table	12× table
× 0 = **0**	8 × 0 = **0**	9 × 0 = **0**	10 × 0 = **0**	11 × 0 = **0**	12 × 0 = **0**
× 1 = **7**	8 × 1 = **8**	9 × 1 = **9**	10 × 1 = **10**	11 × 1 = **11**	12 × 1 = **12**
× 2 = **14**	8 × 2 = **16**	9 × 2 = **18**	10 × 2 = **20**	11 × 2 = **22**	12 × 2 = **24**
× 3 = **21**	8 × 3 = **24**	9 × 3 = **27**	10 × 3 = **30**	11 × 3 = **33**	12 × 3 = **36**
× 4 = **28**	8 × 4 = **32**	9 × 4 = **36**	10 × 4 = **40**	11 × 4 = **44**	12 × 4 = **48**
× 5 = **35**	8 × 5 = **40**	9 × 5 = **45**	10 × 5 = **50**	11 × 5 = **55**	12 × 5 = **60**
× 6 = **42**	8 × 6 = **48**	9 × 6 = **54**	10 × 6 = **60**	11 × 6 = **66**	12 × 6 = **72**
× 7 = **49**	8 × 7 = **56**	9 × 7 = **63**	10 × 7 = **70**	11 × 7 = **77**	12 × 7 = **84**
× 8 = **56**	8 × 8 = **64**	9 × 8 = **72**	10 × 8 = **80**	11 × 8 = **88**	12 × 8 = **96**
× 9 = **63**	8 × 9 = **72**	9 × 9 = **81**	10 × 9 = **90**	11 × 9 = **99**	12 × 9 = **108**
× 10 = **70**	8 × 10 = **80**	9 × 10 = **90**	10 × 10 = **100**	11 × 10 = **110**	12 × 10 = **120**
× 11 = **77**	8 × 11 = **88**	9 × 11 = **99**	10 × 11 = **110**	11 × 11 = **121**	12 × 11 = **132**
× 12 = **84**	8 × 12 = **96**	9 × 12 = **108**	10 × 12 = **120**	11 × 12 = **132**	12 × 12 = **144**

Metric units of length

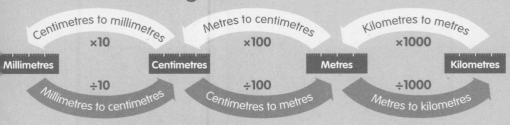

Centimetres to millimetres ×10

Metres to centimetres ×100

Kilometres to metres ×1000

Millimetres   **Centimetres**   **Metres**   **Kilometres**

÷10 Millimetres to centimetres

÷100 Centimetres to metres

÷1000 Metres to kilometres

Metric units of capacity

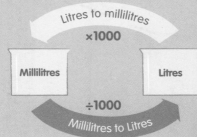

Litres to millilitres ×1000

Millilitres   **Litres**

÷1000 Millilitres to Litres

Metric units of mass

Grams to milligrams ×1000

Kilograms to grams ×1000

Tonnes to kilograms ×1000

milligrams   **Grams**   **Kilograms**   **Tonnes**

÷1000 Milligrams to grams

÷1000 Grams to kilograms

÷1000 Kilograms to tonnes

Formulas for perimeter

Perimeter of a
square = **4a**

Perimeter of a
rectangle = **2 (a + b)**

Perimeter of a
parallelogram
= **2 (a + b)**

Perimeter of an
equilateral triangle
= **3a**

Perimeter of
an isosceles
triangle
= **2a + b**

Perimeter of a
scalene
triangle
= **a + b + c**

Formulas for area

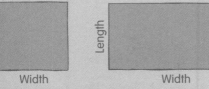
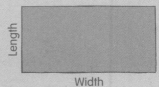
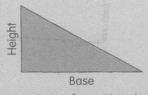

Area of a square
= **width × length**

Area of a rectangle
= **width × length**

Area of a parallelogram
= **base × height**

Area of any triangle
= **½ base × height**

How to be good at maths

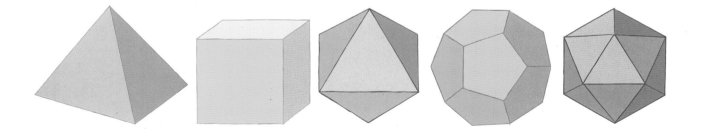

How to be good at maths

Peter Clarke, Caroline Clissold, Cherri Moseley

Editorial consultant Peter Clarke

Senior editor Peter Frances
Senior art editor Mabel Chan

Editors Shaila Brown, Salima Hirani,
Sarah MacLeod, Steve Setford, Rona Skene

Designers Tannishtha Chakraborty,
Louise Dick, Alison Gardner, Mik Gates,
Tessa Jordens, Shahid Mahmood,
Peter Radcliffe, Mary Sandberg, Jacqui Swan,
Steve Woosnam-Savage

Illustrator Acute Graphics

Managing editors Lisa Gillespie, Paula Regan
Managing art editor Owen Peyton Jones

Senior producer, pre-production
Nikoleta Parasaki
Senior producer Mary Slater

Jacket editor Claire Gell
Jacket designers Mark Cavanagh,
Dhirendra Singh
Senior DTP designer Harish Aggarwal
Managing jackets editor Saloni Singh
Design development manager Sophia MTT

Publisher Andrew Macintyre
Art director Karen Self
Design director Phil Ormerod
Publishing director Jonathan Metcalf

First published in Great Britain in 2016 by
Dorling Kindersley Limited
80 Strand, London, WC2R 0RL

Copyright © 2016 Dorling Kindersley Limited
A Penguin Random House Company
10 9 8 7 6 5 4 3 2
007–192676–July/2016

A CIP catalogue record for this book
is available from the British Library.
ISBN: 978-0-2411-8598-8

Printed and bound in China

A WORLD OF IDEAS:
SEE ALL THERE IS TO KNOW

www.dk.com

Contents

3 Measurement

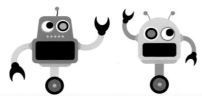

4 Geometry

5 Statistics

6 Algebra

Foreword

Our lives wouldn't be the same without maths. In fact, everything would stop without it. Without numbers we couldn't count a thing, there would be no money, no system of measuring, no shops, no roads, no hospitals, no buildings, no … well, more or less "nothing" as we know it.

For example, without maths we couldn't build houses, forecast tomorrow's weather, or fly a plane. We definitely couldn't send an astronaut into space! If we didn't understand numbers, we wouldn't have TV, the internet, or smartphones. In fact, without numbers, you wouldn't even be reading this book, because it was created on a computer that uses a special number code based on 0s and 1s to store information and make thousands of calculations in a second.

Understanding maths also helps us to understand the world around us. Why do bees make their honeycombs out of hexagons? How can we describe the spiral shape formed by a seashell? Maths holds the answers to these questions and many more.

This book has been written to help you get better at maths, and to learn to love it. You can work through it with the help of an adult, but you can also use it on your own. The numbered steps will talk you through the examples. There are also problems for you to solve yourself. You'll meet some helpful robots, too. They'll give you handy tips and remind you of important mathematical ideas.

Maths is not a subject, it's a language, and it's a universal language. To be able to speak it gives you great power and confidence and a sense of wonder.

Carol Vorderman

2

3 4 5 6

NUMBERS

Numbers are symbols that we use to count and measure things. Although there are just ten number symbols, we can use them to write or count any amount you can think of. Numbers can be positive or negative, and they can be either whole numbers or parts of numbers, called fractions.

Number symbols

Since the earliest times, people have used numbers in their daily lives – to help them count, measure, tell the time, or to buy and sell things.

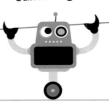

The 10 symbols we use to make up all numbers are called digits.

Number systems

A number system is a set of symbols, called numerals, that represent numbers. Different ancient peoples developed different ways of writing and using numbers.

1 This chart shows the system we use, called the Hindu-Arabic system, compared with some other ancient number systems.

2 Of all these number systems, only ours has a symbol for zero. We can also see that the Babylonian and Egyptian systems are similar.

Numbers were invented to count amounts of things such as apples

	0	1	2	3

Hindu-Arabic numerals are used all over the world today

Many people think the Ancient Egyptian symbols for 1 to 9 represented fingers

	1	2	3
ANCIENT ROMAN	I	II	III
ANCIENT EGYPTIAN			|||
BABYLONIAN	𒑊	𒐀	𒐁

Roman numerals

This chart shows the Roman number system, which puts different letters together to make up numbers.

Symbols after a larger symbol are added to it

Ones	I 1	II 2	III 3	IV 4	V 5	VI 6	VII 7	VIII 8	IX 9
Tens	X 10	XX 20	XXX 30	XL 40	L 50	LX 60	LXX 70	LXXX 80	XC 90
Hundreds	C 100	CC 200	CCC 300	CD 400	D 500	DC 600	DCC 700	DCCC 800	CM 900
Thousands	M 1000	MM 2000	MMM 3000	IV̄ 4000	V̄ 5000	V̄I 6000	V̄II 7000	V̄III 8000	M̄X 9000

1 Look at the symbol for six. It's a V for 5, with I after it, for 1. This means "one more than five", or 5 + 1.

2 Now look at the symbol for nine. This time, the I is before the X. This means "one less than ten" or 10 – 1.

Symbols before a larger symbol are subtracted from it

REAL WORLD MATHS

Zero the hero

Not all number systems have a symbol for zero (0) as we do. On its own, zero stands for "nothing", but when it's part of a bigger number, it's called the place holder. This means it "holds the place" when there is no other digit in that position of a number.

Zeros help us read the time correctly on a 24-hour clock

| 4 | 5 | 6 | 7 | 8 | 9 |

The Babylonian number system is more than 5000 years old

The Romans used letters as symbols for numbers

| IV | V | VI | VII | VIII | IX |

Reading long numbers and dates

To turn a long Roman number or date into a Hindu-Arabic number, we break it into smaller parts then add up the parts.

1 Let's work out the number CMLXXXII. First, we break it into four sections.

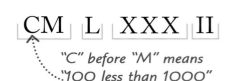

CM │ L │ XXX │ II

"C" before "M" means "100 less than 1000"

2 Next, we work out the values of the different sections. When we add the values together, we get the answer: 982.

$$CM = 1000 - 100 = 900 \ +$$
$$L \qquad\qquad\quad = 50$$
$$XXX = \quad 3 \times 10 = 30$$
$$II = \quad\quad 2 \times 1 = 2$$
$$\overline{\qquad\qquad\qquad\qquad 982}$$

TRY IT OUT

Name the date

Today, we sometimes see dates written in Roman numerals. Can you use what you've learned to work out these years?

1 What's this year?

MCMXCVIII

2 Now have a try at writing these years as Roman numerals:

1666   2015

Answers on page 319

Place value

In our number system, the amount a digit
is worth depends on where it's placed in a
number. This amount is called its place value.

The amount a digit is
worth in a number
is called its place value.

What is place value?

Let's look at the numbers 1, 10, and 100. They are made of the same
digits, 1 and 0, but the digits have different values in each number.

The 10 tens are
exchanged for
one hundred

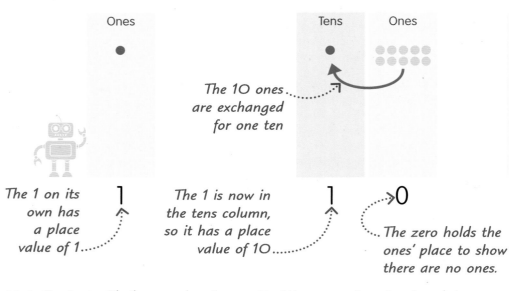

| Ones | | Tens | Ones | | Hundreds | Tens | Ones |

The 10 ones
are exchanged
for one ten

| | 1 | | 1 | 0 | | 1 | 0 | 0 |

The 1 on its
own has
a place
value of 1

The 1 is now in
the tens column,
so it has a place
value of 10

The zero holds the
ones' place to show
there are no ones.

The 1 now has a place
value of 100

1 Let's start with the number 1.
We're going to represent it
by making a ones column and
putting a single dot in it.

2 We can put up to nine dots
in the ones column. When
we get to 10, we exchange the
10 dots in the ones columns for
one in the new tens column.

3 We can show up to 99
using two columns. When
we reach 100, we exchange
the 10 tens for one hundred.

Thousands	H	T	O
	5	7	6

Th	H	T	O
5	0	7	6

4 Now let's put numbers in our columns instead
of dots. We can see that 576 is made up of:
5 groups of 100, or 5 × 100, which is 500
7 groups of 10, or 7 × 10, which is 70
6 groups of 1, or 6 × 1, which is 6.

5 When the number 5067 is put into columns,
we find that the same digits as in Step 4 now
have different place values. For example, the 5 is
now in the thousands column, so its value has
gone up from 500 to 5000.

How place value works

Let's look at the number 2576 and think some more about how place value works.

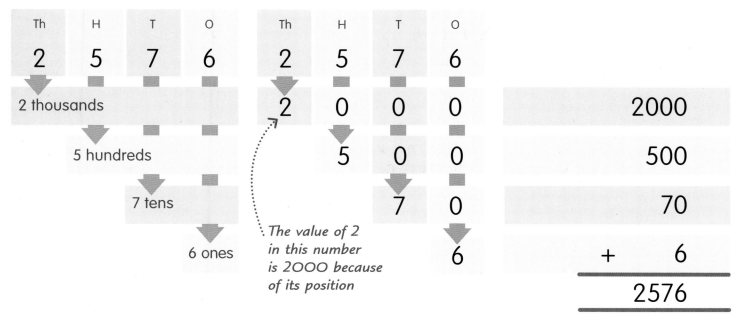

Th	H	T	O
2	5	7	6

2 thousands

5 hundreds

7 tens

6 ones

Th	H	T	O
2	5	7	6
2	0	0	0
	5	0	0
		7	0
			6

The value of 2 in this number is 2000 because of its position

	2000
	500
	70
+	6
	2576

1 When we put the digits into columns, we can see how many thousands, hundreds, tens, and ones the number is made of.

2 When we write this again with numbers, using zeros as place holders, we get four separate numbers.

3 Now, if we add up the four numbers, we get 2576, our original number. So, our place value system works!

Ten times bigger or smaller

Each column in the place-value system increases or decreases the value of a digit by 10. This is really useful when we multiply or divide a number by 10, 100, and so on.

1 Let's look at what happens to 437 when we multiply or divide it by 10.

2 If we divide 437 by 10, each digit moves one column to the right. The new number is 43.7. A dot, called a decimal point, separates ones from numbers 10 times smaller, called tenths.

3 To multiply 437 by 10, we move each digit one column to the left. The new number is 4370, which is 437 × 10.

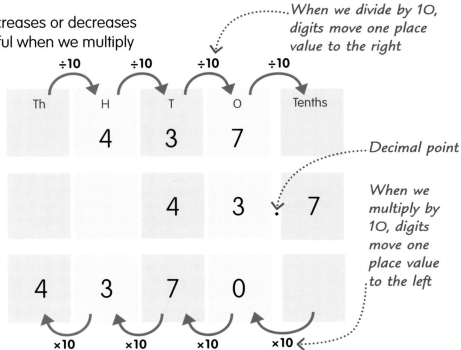

When we divide by 10, digits move one place value to the right

Decimal point

When we multiply by 10, digits move one place value to the left

Sequences and patterns

A sequence is a series of numbers, which we call terms, listed in a special order. A sequence always follows a set pattern, or rule, which means we can work out other terms in the sequence.

A sequence is a set of numbers, called terms, that follow a set pattern, called a rule.

1 Look at this row of houses. The numbers on the doors are 1, 3, 5, and 7. Can we find a pattern in this series?

2 We can see that each number is two more than the one before. So, the rule for this sequence is "add two to each term to find the next term."

3 If we use this rule, we can work out that the next terms are 9 and 11. So, our sequence is: 1, 3, 5, 7, 9, 11, ... The dots show that the sequence carries on.

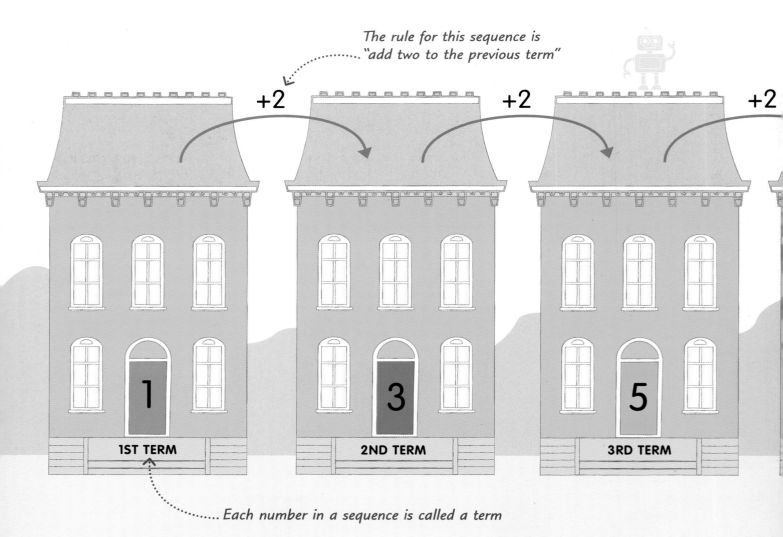

The rule for this sequence is "add two to the previous term"

+2   +2   +2

1
1ST TERM

3
2ND TERM

5
3RD TERM

Each number in a sequence is called a term

Simple sequences

There are lots of ways to make sequences. For example, they can be based on adding, subtracting, multiplying, or dividing.

The dots show that the sequence continues ..

1 In this sequence, we add one to each term to get the next term.

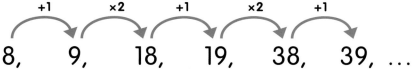

1,   2,   3,   4,   5,   6,   ...

RULE: ADD 1

2 Each term is multiplied by 10 to get the next term in this sequence.

×10   ×10   ×10   ×10

1,   10,   100,   1000,   10 000,   ...

RULE: MULTIPLY BY TEN

3 Sometimes, a rule can have more than one part. In this sequence, we add one, then multiply by two, then go back to adding one, and so on.

+1   ×2   +1   ×2   +1

8,   9,   18,   19,   38,   39, ...

RULE: ADD ONE, THEN MULTIPLY BY TWO

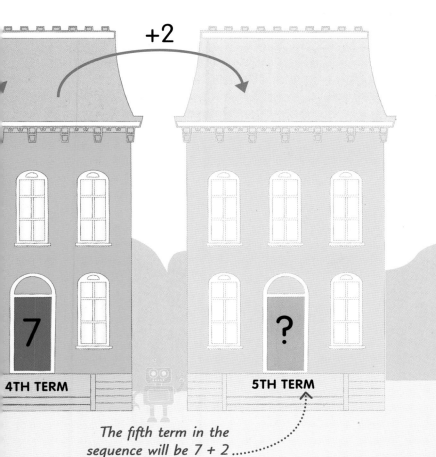

+2

7

4TH TERM

5TH TERM

?

The fifth term in the sequence will be 7 + 2 ....

TRY IT OUT

Spot the sequence

Can you work out the next two terms in each of these sequences? You'll have to work out the rule for each sequence first – a number line might help you.

1 22, 31, 40, 49, 58, ...

2 4, 8, 12, 16, 20, ...

3 100, 98, 96, 94, ...

4 90, 75, 60, 45, 30, ...

Answers on page 319

Sequences and shapes

Some number sequences can be used to create shapes by using the terms in the sequence to measure the parts of a shape, such as the lengths of its sides.

Triangular numbers

One sequence that can be shown as shapes is the triangular number sequence. If we take a whole number and add it to all the other whole numbers that are less than that number, we get this sequence: 1, 3, 6, 10, 15, … Each of the numbers can be shown as a triangle.

We can show the triangular sequence by using shapes

1 The sequence starts with 1, shown as a single shape.

2 When we add 2, we can arrange the shapes in a triangle.
1 + 2 = 3

Each new number adds a new row to the triangle's base

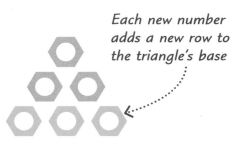

3 Adding 3 makes a new triangle.
1 + 2 + 3 = 6

4 Now we add 4 to make a fourth triangle.
1 + 2 + 3 + 4 = 10

5 Adding 5 creates a fifth triangle, and so on.
1 + 2 + 3 + 4 + 5 = 15

Square numbers

If we multiply each of the numbers 1, 2, 3, 4, 5 by themselves, we get this sequence: 1, 4, 9, 16, 25, ...
We can show this number sequence as real squares.

The fourth square number is 16

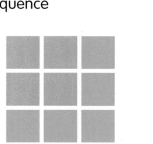

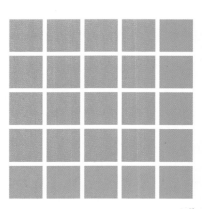

1 × 1 = 1   **2 × 2 = 4**   **3 × 3 = 9**   **4 × 4 = 16**   **5 × 5 = 25**

Pentagonal numbers

The sides of these five-sided shapes, called pentagons, are made up of equally spaced dots. If we start with one dot, and then count the dots in each pentagon, we see this sequence: 1, 5, 12, 22, 35, ... These numbers are called pentagonal numbers.

Each pentagon shares one corner, called a vertex, with the other pentagons

Each pentagon has five sides with equal numbers of dots

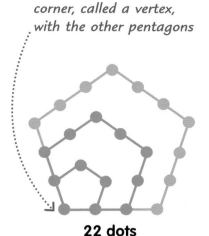

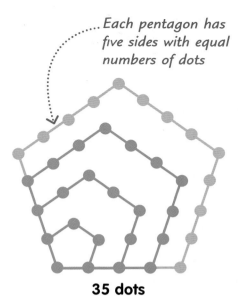

1 dot **5 dots** **12 dots** **22 dots** **35 dots**

REAL WORLD MATHS

The Fibonacci sequence

One of the most interesting sequences in maths is the Fibonacci sequence, named after a 13th-century Italian mathematician. The first two terms of the sequence are 1. Then we add the two previous terms together to get the next term.

Sequence starts at 1......

Add the previous two terms to find the next term

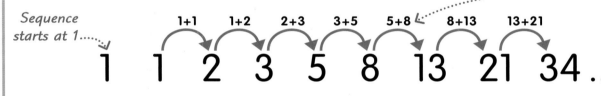

1+1   1+2   2+3   3+5   5+8   8+13   13+21

1   1   2   3   5   8   13   21   34 ...

We can use the number sequence to make a pattern of boxes like this

When we connect the boxes' opposite corners, we draw a spiral shape

34   **21**   **13**   **8**   **5**   **3   2   1**

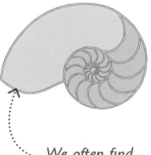

We often find Fibonacci spirals, like this shell, in nature

Positive and negative numbers

Positive numbers are all the numbers that are greater than zero. Negative numbers are less than zero, and they always have a negative sign (–) in front of them.

> Negative numbers have a '–' before them. Positive numbers usually have no sign in front of them.

What are positive and negative numbers?

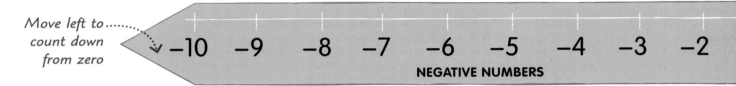

Move left to count down from zero

–10   –9   –8   –7   –6   –5   –4   –3   –2

NEGATIVE NUMBERS

1 If we put numbers on a line called a number line, like the line on this signpost, we see that negative numbers count back from zero, while positive numbers get larger from the zero point.

2 Negative numbers are numbers less than zero. In calculations, we put negative numbers in brackets, like this (–2), to make them easier to read.

Adding and subtracting positive and negative numbers

Here are some simple rules to remember when we add and subtract positive and negative numbers. We can show how they work on a simple version of our numbers signpost, called a number line.

1 **Adding a positive number**
When we add a positive number, we move to the right on the number line.
2 + 3 = 5

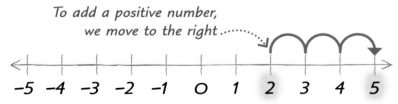

To add a positive number, we move to the right

–5 –4 –3 –2 –1   O   1   2   3   4   5

2 **Subtracting a negative number**
To subtract a negative number, we also move right on the number line. So, subtracting –3 from 2 is the same as 2 + 3.
2 – (–3) = 5

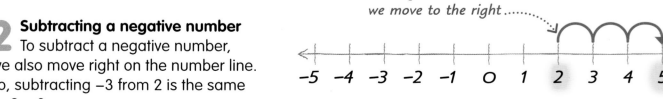

To subtract a negative number, we move to the right

–5 –4 –3 –2 –1   O   1   2   3   4   5

REAL WORLD MATHS

Ups and downs

We sometimes use positive and negative numbers to describe the floors in a building. Floors below ground level often have negative numbers.

TRY IT OUT

Positively puzzling

Use a number line to work out these calculations.

1 $7 - (-3) = ?$ **3** $7 + (-9) = ?$

2 $-4 + (-1) = ?$ **4** $-2 - (-7) = ?$

Answers on page 319

Move to the right to count up from zero

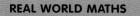

0 1 2 3 4 5 6 7 8 9 10

POSITIVE NUMBERS

3 Zero (0) is not positive or negative. It's the separation point between the positive and negative numbers.

4 We don't usually put any sign in front of positive numbers. So, when you see a number without a sign, it's always positive.

To subtract a positive number, move to the left on the number line

3 Subtracting a positive number
Now let's try subtracting a positive number. To subtract 3 from 2, we move to the left to get the answer.
2 − 3 = −1

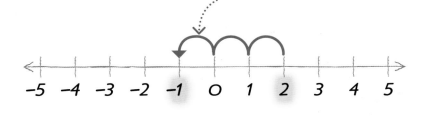

To add a negative number, move to the left on the number line

4 Adding a negative number
When we add a negative number, it gives the same answer as subtracting a positive one. To add −3 to 2, we move left on the number line.
2 + (−3) = −1

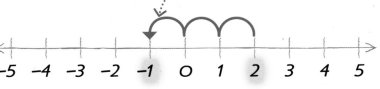

Comparing numbers

We often need to know if a number is the same as, smaller than, or larger than another number. We call this comparing numbers.

We use comparison symbols to show the relationship between two numbers.

More, less, or the same?

When we compare amounts in everyday life, we use words like more, less, larger, smaller, or the same as. In maths, we say numbers or amounts are greater than, less than, or equal to each other.

1 Equal
Look at this tray of cupcakes. There are five cakes in each row. So, the number in one row is equal to the number in the other.

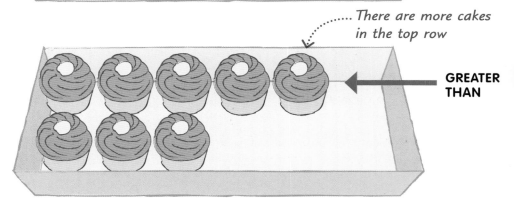

The number of cakes in each row is the same

EQUAL

2 Greater than
Now there are five cakes in the top row and three in the bottom one. So, the number in the top row is greater than the number in the bottom one.

There are more cakes in the top row

GREATER THAN

3 Less than
This time, there are five cakes in the top row and six in the bottom row. So, the number in the top row is less than the number in the bottom.

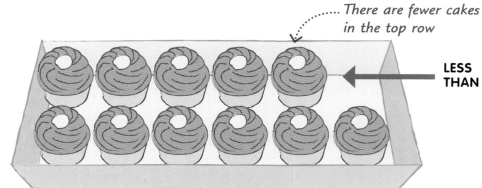

There are fewer cakes in the top row

LESS THAN

Using symbols to compare numbers

We use these signs, called comparison symbols, when we compare numbers or amounts.

.. *The narrowest part of the symbol points to the smaller number*

1 Equals
This symbol means "is equal to".
For example, 90 + 40 = 130 means "90 + 40 is equal to 130".

2 Greater than
This symbol means "is greater than".
For example, 24 > 14 means "24 is greater than 14".

3 Less than
This symbol means "is less than".
For example, 11 < 32 means "11 is less than 32".

Significant digits

The significant digits of a number are the digits that influence the value of the number. When we compare numbers, significant digits are very useful.

1 This number has four digits. The most significant digit is the one with the highest place value, and so on, down to the least significant digit.

Most significant digit

Third significant digit

1 4 0 4

Second significant digit

Least significant digit

2 Let's compare 1404 and 1133. The place value of the most significant digits is the same, so we compare the second most significant digits.

Th	H	T	O
1	4	0	4
1	1	3	3

...*The most significant digits are the same*

3 The second most significant digit of 1404 is larger than it is in 1133. So, 1404 is the larger number.

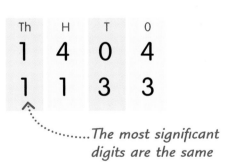

1 4 0 4 > 1 1 3 3

...*The second significant digit is larger in this number*

TRY IT OUT

Which symbol?

Complete each of these examples by adding one of the three symbols you've learned.

Here's a reminder of the three symbols you'll need:

— Equals

> Is greater than

< Is less than

1 5123 ? 10 221

2 −2 ? 3

3 71 399 ? 71 100

4 20 − 5 ? 11 + 4

Answers on page 319

Ordering numbers

Sometimes we need to compare a whole series of numbers so that we can put them in order. To do this, we use what we know about place values and significant figures.

Xoon
912
votes

1 Cybertown has held an election for mayor. We need to put the candidates in order of the votes they received.

	TTh	Th	H	T	O
Xoon			9	1	2
Zeet				4	5
Moop		5	2	3	4
Flug			4	4	4
Krog	1	0	4	2	3
Jeek		5	1	2	1

2 First, we put the candidates' votes into a table so we can compare the place value of their most significant digits.

	TTh	Th	H	T	O
Krog	1	0	4	2	3

The first significant figure is the one furthest to the left

3 Let's look at the most significant digits. Only Krog's total has a digit in the ten thousands column. So, his vote total is the highest and we can put it first in a new table.

	TTh	Th	H	T	O
Krog	1	0	4	2	3
Moop		5	2	3	4
Jeek		5	1	2	1

VOTE KROG!

	TTh	Th	H	T	O
Krog	1	0	4	2	3
Moop		5	2	3	4
Jeek		5	1	2	1
Xoon			9	1	2
Flug			4	4	4
Zeet				4	5

4 When we compare second significant digits, we see Moop and Jeek have the same digit in the thousands. So, we compare third significant digits. Moop's digit is greater than Jeek's.

5 We carry on comparing digits in the place-value columns until we have put the whole list in order, from largest to smallest numbers. Krog is the new mayor!

Zeet 45 votes

Moop 5234 votes

Flug 444 votes

Krog 10 423 votes

Jeek 5121 votes

Ascending and descending order

When we put things in order, sometimes we want to put the largest number first, and sometimes the smallest.

1 In a maths test, there were 100 questions. Amira got 94 correct; Bella got 45; Claudia got 61; Danny got 35; Ethan got 98; Fiona got 31; Greta got 70; and Harry got 81.

2 When the scores are listed from the highest to the lowest, we call it descending order.

3 When we order the scores from lowest to highest, we call it ascending order.

DESCENDING ORDER
98
94
81
70
61
45
35
31

ASCENDING ORDER
31
35
45
61
70
81
94
98

TRY IT OUT

All in order

Practise your ordering skills by putting this list of ages in ascending order. Why not make an ordered list based on your own friends and family? You could order them by age, height, or the day of the month of their birthday.

Answer on page 319

NAME	AGE
Jake (me!)	9
Mum	37
Trevor the gerbil	1
Dad	40
Grandpa	67
Buster the dog	7
Grandma	68
Uncle Dan	35
Anna (my sister)	13
Bella the cat	3

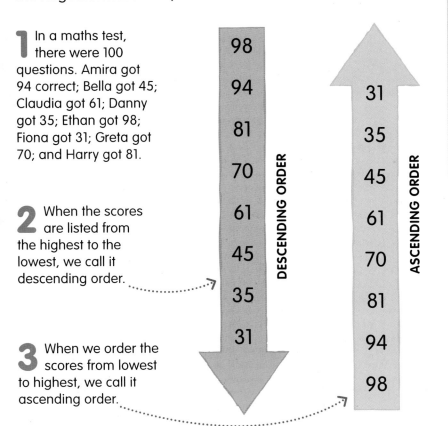

Estimating

Sometimes when we're measuring or calculating, we don't need to work out the exact answer – a sensible guess, called an estimate, is good enough.

Estimation is finding something that is close to the correct answer.

Approximately equal

1 Equal
We've already learned the symbol to use for things that are equal.

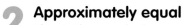

2 Approximately equal
This is the symbol we use for things that are nearly the same. In maths, we say they are approximately equal.

Quick counting

In everyday life, we often don't need to count something exactly. It's enough to have a good idea of how many things there are or roughly how big something is.

Compare the baskets to estimate which one has the most strawberries in it

1 These three baskets of strawberries all cost the same, but they contain different numbers of strawberries.

2 We don't actually have to count to see that the third basket contains more strawberries than the other two. So, the third basket is the best bargain.

Any basket for £1

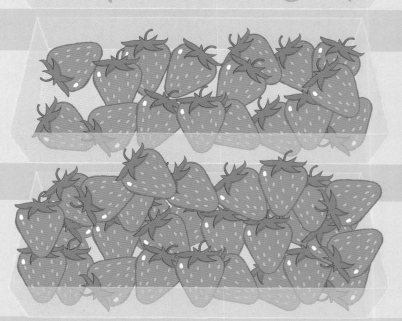

Estimating a total

Sometimes we estimate because it would take too long to count or calculate the exact answer.

1 Let's look at this bed of tulips. We want to know roughly how many there are, without having to count them, one by one.

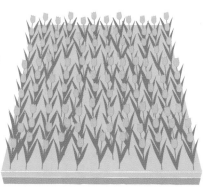

There are nine horizontal rows

2 The tulips aren't in exact rows, but we can count 11 flowers in the front row. There are nine rows, so we can say there are about 11 × 9 flowers, which is 99.

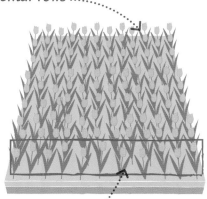

There are 11 flowers in the front row

The flower bed is divided roughly into nine squares

3 Another way to estimate the total is to divide the bed into rough squares. If we count the flowers in one square, we can estimate the number in the whole bed.

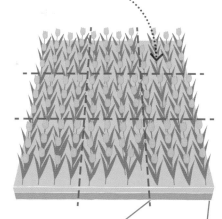

4 There are 12 tulips in the bottom right square. So, the total number is approximately 12 × 9, which is 108.

There are 12 flowers in the bottom right square

5 Our two estimates have come up with answers of 99 and 108. In fact, there are 105 tulips, so both estimates were pretty close!

Checking a calculation

Sometimes, we work out what we expect an answer to be by simplifying, or rounding, the numbers.

We estimate that the answer will be approximately 7000

$$2847 + 4102 = \boxed{?}$$ $$3000 + 4000 = \boxed{7000}$$ $$2847 + 4102 = \boxed{6949}$$

1 Let's add together 2847 and 4102. We make an estimate first so that if our answer is very different, we know that we might have made a mistake.

2 The first number is slightly less than 3000, and the second is slightly more than 4000. We can quickly add 3000 to 4000, to get 7000.

3 When we do the actual calculation, the answer we get is very close to our estimate. So, we can be confident that our addition is correct.

Rounding

Rounding means changing a number to another number that is close to it in value, but is easier to work with or remember.

The rounding rule is that for digits less than 5, we round down. For digits of 5 or more, we round up.

Rounding up and rounding down

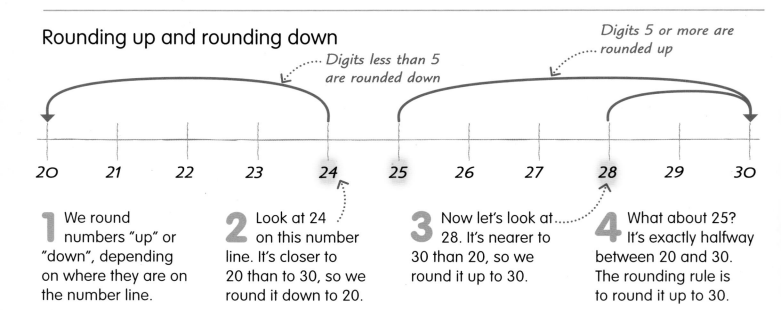

Digits less than 5 are rounded down

Digits 5 or more are rounded up

1 We round numbers "up" or "down", depending on where they are on the number line.

2 Look at 24 on this number line. It's closer to 20 than to 30, so we round it down to 20.

3 Now let's look at 28. It's nearer to 30 than 20, so we round it up to 30.

4 What about 25? It's exactly halfway between 20 and 30. The rounding rule is to round it up to 30.

Rounding using place value

When we round numbers, we use the place values of a number's digits.

1 **Rounding to the nearest ten**
We use the ones digit to decide whether to round up or down to the nearest ten. Let's round 83 and 89.

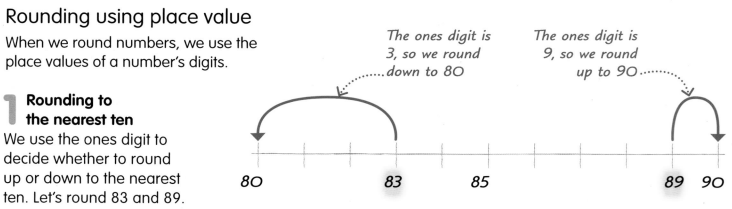

The ones digit is 3, so we round down to 80

The ones digit is 9, so we round up to 90

2 **Rounding to the nearest hundred**
To round to the nearest 100, we look at the tens digit and follow the rounding rule. Let's round 337 and 572.

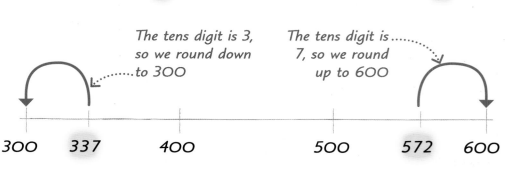

The tens digit is 3, so we round down to 300

The tens digit is 7, so we round up to 600

Rounding to different place values

Rounding to different place values will give us different results. Let's look at what happens to 7641 when we round it to different place values.

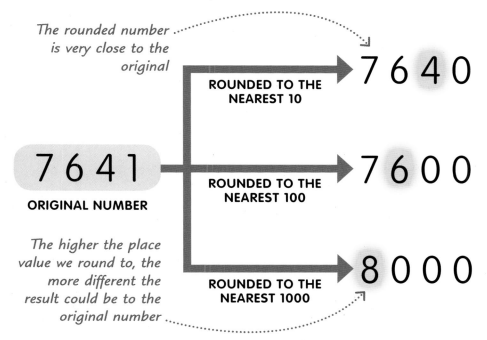

The rounded number is very close to the original

7641
ORIGINAL NUMBER

ROUNDED TO THE
NEAREST 10
7640

ROUNDED TO THE
NEAREST 100
7600

The higher the place value we round to, the more different the result could be to the original number

ROUNDED TO THE
NEAREST 1000
8000

TRY IT OUT

Estimating height

This robot is 165 cm tall.

1 What is his height rounded to the nearest 10 cm?

2 What is his height rounded to one significant digit? (See below.)

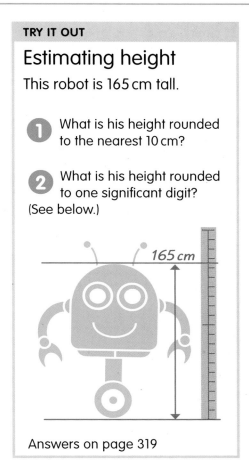

165 cm

Answers on page 319

Rounding to significant digits

We can also round numbers to one or more significant digits.

1 Let's look at the number 6346. The most significant digit is the one with the highest place value. So, 6 is the most significant digit. The digit after it is less than 5, so we round down to 6000.

2 The second significant digit is in the hundreds. The next digit is less than 5, so when we round to two significant digits, 6346 becomes 6300.

3 The third significant digit is in the tens column. If we round our number to three significant digits, it becomes 6350.

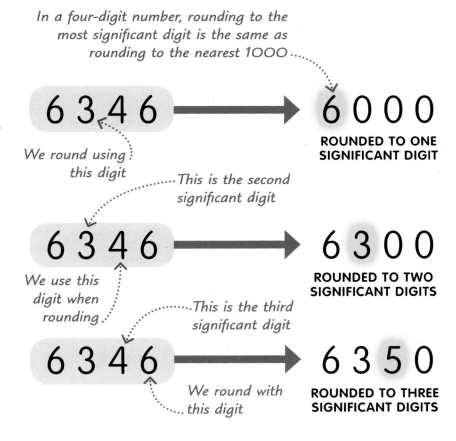

In a four-digit number, rounding to the most significant digit is the same as rounding to the nearest 1000

6346

We round using this digit

6000
ROUNDED TO ONE
SIGNIFICANT DIGIT

This is the second significant digit

6346

We use this digit when rounding

6300
ROUNDED TO TWO
SIGNIFICANT DIGITS

This is the third significant digit

6346

We round with this digit

6350
ROUNDED TO THREE
SIGNIFICANT DIGITS

Factors

A factor is a whole number that can be divided or shared into another number.
Every number has at least two factors, because it can be divided by itself and 1.

What is a factor?

This chocolate bar is made up of 12 squares. We can use it to find the factors
of 12 by working out how many ways we can share it into equal parts.

$12 \div 1 = 12$

1 If we divide the 12-square bar by one, it stays whole. So, 1 and 12 are both factors of 12.

$12 \div 2 = 6$

2 Dividing the bar into two gives two groups of six squares. So, 2 and 6 are also factors of 12.

$12 \div 3 = 4$

3 When we divide the bar into three, we get three groups of four. So, 3 and 4 are factors of 12.

$12 \div 4 = 3$

4 When the bar's divided into four, we get four groups of three squares. We already know that 4 and 3 are factors of 12.

$12 \div 6 = 2$

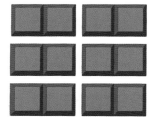

5 Dividing the bar by six gives six groups of two squares. We have already found that 6 and 2 are factors of 12.

$12 \div 12 = 1$

6 Finally, we can divide the bar into 12 and get 12 groups of one square. We've now found all the factors of 12.

Factor pairs

Factors always come in pairs. Two numbers that make a new number when multiplied together are called a factor pair.

$1 \times 12 = 12$ or $12 \times 1 = 12$

$2 \times 6 = 12$ or $6 \times 2 = 12$

$3 \times 4 = 12$ or $4 \times 3 = 12$

1 Let's look again at the factors of 12 we found. Each pair can be written in two different ways.

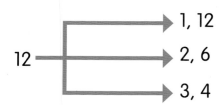

2 So, the factor pairs of 12, written in either order, are: 1 and 12, 2 and 6, and 3 and 4.

Finding all the factors

If you need to find all the factors of a number, here's a way to write down your findings to make sure you don't miss any out.

1 To find all the factors of 30, first write 1 at the beginning of a line and 30 at the other end, because we know that every number has 1 and itself as factors.

$$1 \times 30 = 30$$

| 1 | | 30 |

2 Next, we test whether 2 is a factor and find that 2 × 15 = 30. So, 2 and 15 are factors of 30. We put 2 just after 1 and 15 at the other end, just before 30.

$$2 \times 15 = 30$$

1   2   15   30

3 Next, we check 3 and find that 3 × 10 = 30. So, we can add 3 and 10 to our row of factors, the 3 after 2 and the 10 before 15.

$$3 \times 10 = 30$$

1   2   3   10   15   30

4 When we check 4, we can't multiply it by another whole number to make 30. So, 4 isn't a factor of 30. It doesn't go on our line.

$$4 \times ? = 30$$

1   2   3   10   15   30

5 We check 5 and find that 5 × 6 = 30. So we add 5 after 3, and 6 before 10. We don't need to check 6 because it's already on our list. So, our row of factors of 30 is complete.

$$5 \times 6 = 30$$

1   2   3   5   6   10   15   30

Common factors

When two or more numbers have the same factors, we call them common factors.

1 Here are the factors of 24 and 32. Both have factors of 1, 2, 4, and 8. These are their common factors, in yellow circles.

2 The largest of the common factors is 8. We call it the highest common factor, sometimes shortened to HCF.

The highest common factor is 8

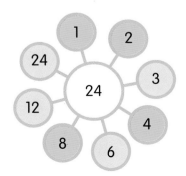

FACTORS OF 24   **FACTORS OF 32**

Multiples

When two whole numbers are multiplied together,
we call the result a multiple of the two numbers.

A multiple of a number is
that number multiplied by
any other whole number.

Finding multiples

*The number 12 is a
multiple of both 3 and 4*

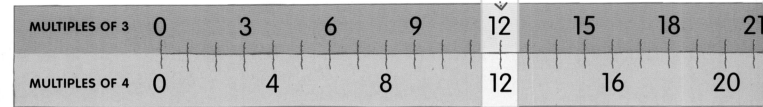

| MULTIPLES OF 3 | 0 | 3 | 6 | 9 | 12 | 15 | 18 | 21 |
| MULTIPLES OF 4 | 0 | 4 | 8 | | 12 | 16 | | 20 |

1 We can use a number line like this to work out a number's multiples. And if you know your multiplication tables, you'll find working with multiples is even easier!

2 Above the line we have marked the first 16 multiples of 3. To find the multiples, we multiply 3 by 1, then 2, then 3, and so on: $3 \times 1 = 3$, $3 \times 2 = 6$, $3 \times 3 = 9$

Common multiples

We have found out that some numbers can be multiples of more than one number. We call these common multiples.

1 This is a Venn diagram. It's another way of showing the information in the number line above. In the blue circle are multiples of 3 from 1 to 50. The green circle shows all the multiples of 4 from 1 to 50.

2 There are four numbers in the section where the circles overlap: 12, 24, 36, and 48. These are the common multiples of 3 and 4.

3 The lowest common multiple of 3 and 4 is 12. We don't know their highest common multiple, because numbers can be infinitely large.

*We call the smallest number
in the overlapping section
the lowest common multiple*

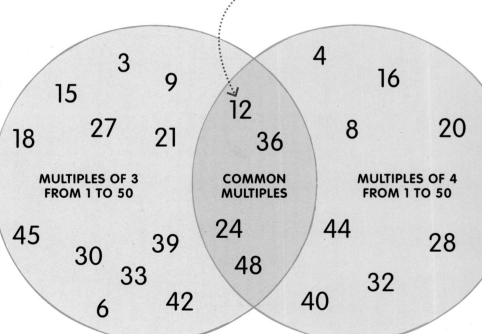

MULTIPLES OF 3
FROM 1 TO 50

COMMON
MULTIPLES

MULTIPLES OF 4
FROM 1 TO 50

TRY IT OUT

Multiple mayhem

Which numbers are multiples of 8 and which are multiples of 9? Can you find any common multiples of 8 and 9?

Answers on page 319

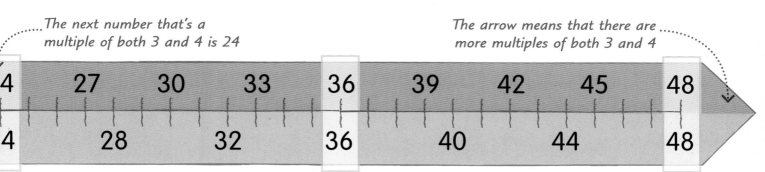

64 **32** 36 **48**
16 81 108 56 90
72 144 27 18

The next number that's a multiple of both 3 and 4 is 24

The arrow means that there are more multiples of both 3 and 4

| 24 | 27 | 30 | 33 | 36 | 39 | 42 | 45 | 48 |

| 24 | 28 | 32 | 36 | 40 | 44 | 48 |

3 Multiples of 4 are marked below the number line. Look at the number 12. It appears on both lines. So it's a multiple of both 3 and 4.

4 Multiples and factors work together – we multiply two factors together to get a multiple. So 3 and 4 are factors of 12, and 12 is a multiple of 3 and 4.

Finding the lowest common multiple

Here's a way of finding the lowest common multiple of three numbers.

1 Let's find the lowest common multiple of 2, 4, and 6. First, we draw a number line showing the first ten multiples of 2.

2 Now we draw a number line showing the multiples of 4. We find that 4, 8, 12, 16, and 20 are common multiples of 2 and 4.

3 When we draw a number line of the multiples of 6, we see that the first common multiple of all three numbers is 12. So 12 is the lowest common multiple of 2, 4, and 6.

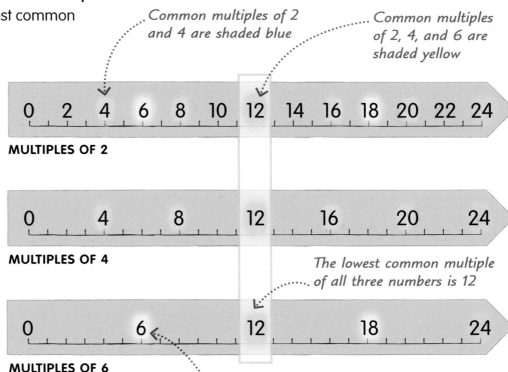

Common multiples of 2 and 4 are shaded blue

Common multiples of 2, 4, and 6 are shaded yellow

0 2 4 6 8 10 12 14 16 18 20 22 24
MULTIPLES OF 2

0 4 8 12 16 20 24
MULTIPLES OF 4

The lowest common multiple of all three numbers is 12

0 6 12 18 24
MULTIPLES OF 6

Common multiples of 2 and 6 are shaded white

Prime numbers

A prime number is a whole number greater than 1 that can't be divided by another whole number except for itself and 1.

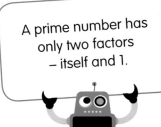

A prime number has only two factors – itself and 1.

Finding prime numbers

To find out whether or not a number is prime, we can try to divide it exactly by other whole numbers. Let's try this out on a few numbers.

1 Is 2 a prime number?
We can divide 2 by 1 and also by itself. But we can't divide 2 by any other number. So, we know 2 is a prime number.

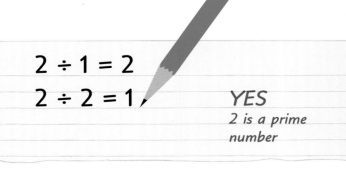

$2 \div 1 = 2$
$2 \div 2 = 1$

YES
2 is a prime number

2 Is 4 a prime number?
We can divide 4 by 1 and by itself. Can we divide 4 exactly by any other number? Let's try dividing by 2: $4 \div 2 = 2$
We can divide 4 by 2, so 4 is not a prime number.

$4 \div 1 = 2$
$4 \div 4 = 1$
$4 \div 2 = 2$

NO
4 is not a prime number

3 Is 7 a prime number?
We can divide 7 by 1 and by itself. Now let's try dividing 7 by other numbers. We can't divide 7 exactly by 2, 3, or 4. We can stop checking once we get over half of the number we're looking at – in this example, once we get to 4. So, 7 is a prime number.

$7 \div 1 = 7$
$7 \div 7 = 1$

YES
7 is a prime number

4 Is 9 a prime number?
We can divide 9 by 1 and by itself. We can't divide 9 exactly by 2, but we can divide it by 3: $9 \div 3 = 3$
This means 9 is not a prime number.

$9 \div 1 = 9$
$9 \div 9 = 1$
$9 \div 3 = 3$

NO
9 is not a prime number

Prime numbers up to 100

This table shows all the prime numbers from 1 to 100.

1	2	3	4	5	6	7	8	9	10
11	12	13	14	15	16	17	18	19	20
21	22	23	24	25	26	27	28	29	30
31	32	33	34	35	36	37	38	39	40
41	42	43	44	45	46	47	48	49	50
51	52	53	54	55	56	57	58	59	60
61	62	63	64	65	66	67	68	69	70
71	72	73	74	75	76	77	78	79	80
81	82	83	84	85	86	87	88	89	90
91	92	93	94	95	96	97	98	99	100

1 is not a prime number because it doesn't have two different factors — 1 and itself are the same number!

2 is the only even prime. All other even numbers can be divided by 2 so are not prime

Prime numbers are shaded dark purple

Non-primes are shaded pale purple

Prime or not prime?

There's a simple trick we can use to check whether a number is prime – just follow the steps on this chart:

PICK A WHOLE NUMBER FROM 2 TO 100

CAN YOU DIVIDE THIS NUMBER EXACTLY BY 2, 3, 5, OR 7?

NO

YES

IT'S A PRIME

IT'S NOT A PRIME

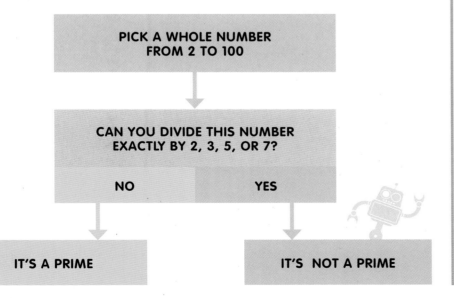

REAL WORLD MATHS

The largest prime

The ancient Greek mathematician Euclid worked out that we can never know the largest possible prime number. The largest prime we currently know is more than 22 million digits long! It's written like this:

$$2^{74\,207\,281} -1$$

This means "multiply 2 by itself 74 207 281 times, then subtract 1"

Prime factors

A factor of a whole number that is also a prime number is called a prime factor.
One of the special things about prime numbers is that any whole number is either
a prime number or can be found by multiplying two or more prime factors.

Finding prime factors

Prime numbers are like the building blocks of numbers, because
every number that's not a prime can be broken down into
prime factors. Let's find the prime factors of 30.

*Prime factors have a
green circle round them*

$$30 \div ②= 15$$

2 and 15 are factors of 30

$$15 \div 2 = ?$$

2 is not a factor of 15

1 We start by seeing if we can divide 30 by 2,
the smallest prime number. We can divide 30
exactly by 2, and 2 is a prime number, so we can
say 2 is one of 30's prime factors.

2 Now let's look at 15, the factor pair of 2 in the
last step. It's not a prime number, so we have
to break it down more. We can't divide it exactly by
2, so let's try another number.

$$15 \div ③ = ⑤$$

3 and 5 are factors of 15

$$30 = ② \times ③ \times ⑤$$

2, 3, and 5 are prime factors of 30

3 We can divide 15 exactly by 3 and get 5. Both
3 and 5 are prime numbers, so they must also
be prime factors of 30.

4 So, we can say that 30 is the product of
multiplying together three prime factors –
2, 3, and 5.

REAL WORLD MATHS

Prime factors for internet security

When we send information over the internet, it's turned
into code to keep it secure. These codes are based on
prime factors of very large numbers, which fraudsters
would find really difficult and time-consuming to find.

53528154482532
53528154482532
53528154482532
53528154482532
53528154482532
53528154482532
53528154482532
53528154482532

All whole numbers
can be broken down
into two or more
prime factors.

Factor trees

An easy way to find the prime factors of a number is to draw a diagram called a factor tree.

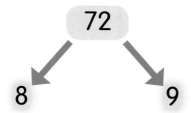

1 Let's find the prime factors of 72. We know from our multiplication tables that 8 and 9 are factors of 72, so we can write the information like this.

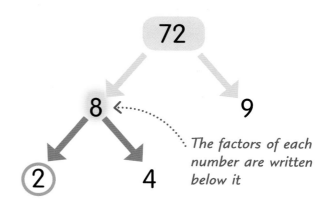

The factors of each number are written below it

2 Neither 8 or 9 are prime numbers, so we need to break them down some more. When we factor 8, we get 2 and 4. We put a circle round 2, because it's a prime number.

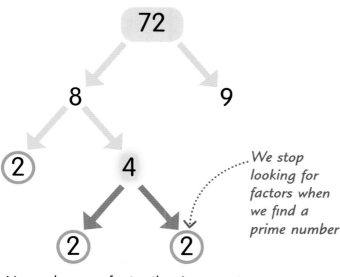

We stop looking for factors when we find a prime number

3 Now when we factor the 4, we get 2 and 2. Both are prime numbers so we circle them, too.

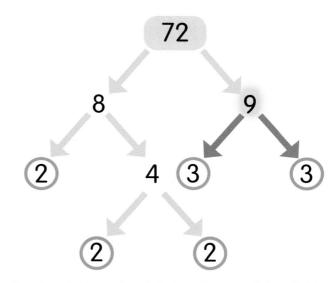

4 Now let's go back to the 9. It can't be divided by 2, but it can be divided by 3, giving two factors of 3. Both are prime numbers, so now we can write all the prime factors of 72 like this:
72 = 2 × 2 × 2 × 3 × 3

TRY IT OUT

Different tree, same answer

There are often lots of ways to make a factor tree. Here's another tree for 72, starting by dividing it by 2. Can you finish it? There's more than one way – as long as you get the same list of prime factors as in Step 4, you've done it correctly!

Answer on page 319

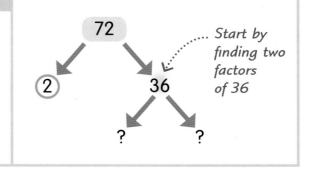

Start by finding two factors of 36

Square numbers

When we multiply a whole number by itself, the result is a square number. Square numbers have a special symbol, a small "2" after the number, like this: 3^2.

The square measures 2 × 2 small squares

$2 \times 2 = 4$ or $2^2 = 4$

1	2
3	4

1 We can show the squares of numbers as actual squares. So to show 2^2, we can make a square that's made up of four smaller squares. So, 4 is a square number.

$3 \times 3 = 9$ or $3^2 = 9$

1	2	3
4	5	6
7	8	9

2 To show 3^2, our new square is three squares wide and three squares deep – a total of nine squares. This means 9 is also a square number.

$4 \times 4 = 16$ or $4^2 = 16$

1	2	3	4
5	6	7	8
9	10	11	12
13	14	15	16

3 When we show 4^2 as a square, it's made of 4 × 4 small squares, which makes a total of 16 squares.

$5 \times 5 = 25$ or $5^2 = 25$

1	2	3	4	5
6	7	8	9	10
11	12	13	14	15
16	17	18	19	20
21	22	23	24	25

4 This is 5^2 shown as 5 × 5 squares. There are 25 squares, which is the same as 5 multiplied by 5. So, the four square numbers after 1 are 4, 9, 16, and 25.

Squares table

1 This table shows the squares of numbers up to 12 × 12. Let's see how it works by finding the square of 7. First, find 7 on the top row. ....

The square numbers form a diagonal line within the grid

×	1	2	3	4	5	6	7	8	9	10	11	12
1	1	2	3	4	5	6	7	8	9	10	11	12
2	2	4	6	8	10	12	14	16	18	20	22	24
3	3	6	9	12	15	18	21	24	27	30	33	36
4	4	8	12	16	20	24	28	32	36	40	44	48
5	5	10	15	20	25	30	35	40	45	50	55	60
6	6	12	18	24	30	36	42	48	54	60	66	72
7	7	14	21	28	35	42	49	56	63	70	77	84
8	8	16	24	32	40	48	56	64	72	80	88	96
9	9	18	27	36	45	54	63	72	81	90	99	108
10	10	20	30	40	50	60	70	80	90	100	110	120
11	11	22	33	44	55	66	77	88	99	110	121	132
12	12	24	36	48	60	72	84	96	108	120	132	144

2 Now find 7 in the left-hand column. Follow the row and column until you get to the square where they meet. This square contains the square of that number.

3 The row and column meet at the square containing 49. So, the square of 7 is 49.

Squares of odd numbers are always odd

Squares of even numbers are always even

Square roots

A square root is a number that you multiply by itself once to get a particular square number. The symbol we use for the square root is √.

Square roots are the opposite, or inverse, of square numbers.

1 Let's look at 36. Its square root is 6, the number that we multiply by itself, or square, to get 36. We write it like this: √36 = 6

$$\sqrt{36} = 6$$

because

$$6 \times 6 = 36 \text{ or } 6^2 = 36$$

2 Squares and square roots are opposites – so if 25 is the square of 5, then 5 is the square root of 25. The word we use in maths for this is "inverse".

Square

We square 5 to get 25

5   25

5 is the square root of 25

Square root

3 We can use this squares table to find square roots. Let's look at the square number 64. To find its square root, follow its row and column back to the start. We find 8 at the start of 64's row and column, so we know 8 is the square root of 64.

×	1	2	3	4	5	6	7	8	9	10	11	12
1	1	2	3	4	5	6	7	8	9	10	11	12
2	2	4	6	8	10	12	14	16	18	20	22	24
3	3	6	9	12	15	18	21	24	27	30	33	36
4	4	8	12	16	20	24	28	32	36	40	44	48
5	5	10	15	20	25	30	35	40	45	50	55	60
6	6	12	18	24	30	36	42	48	54	60	66	72
7	7	14	21	28	35	42	49	56	63	70	77	84
8	8	16	24	32	40	48	56	64	72	80	88	96
9	9	18	27	36	45	54	63	72	81	90	99	108
10	10	20	30	40	50	60	70	80	90	100	110	120
11	11	22	33	44	55	66	77	88	99	110	121	132
12	12	24	36	48	60	72	84	96	108	120	132	144

Follow the row or column back to find the square root

The square numbers are in dark purple

TRY IT OUT

Find the roots

Use the table on this page to work out the answers to these questions.

1 10 is the square root of which number?

2 4 is the square root of which number?

3 What is the square root of 81?

Answers on page 319

Cube numbers

A cube number is the result of multiplying a number by itself, and then by itself again.

How to cube a number

$$2 \times 2 \times 2 = ?$$

$$2 \times 2 = 4$$
$$4 \times 2 = 8$$

1 Let's find the cube of 2. First, we multiply 2×2 to get 4. Then we multiply the answer, 4, by 2 again to make 8.

$$2^3 = 8$$

because

$$2 \times 2 \times 2 = 8$$

2 So, now we know that the cube of 2 is 8. When we cube numbers, we use a special symbol – a small "3" after the number, like this: 2^3.

Cube number sequence

Each cube number can be shown by an actual cube, made from cubes of one unit.

1 Let's start with 1: $1^3 = 1$.
We can show the cube number as a single cube, like this.

All the cube's sides are one unit long

$$1 \times 1 \times 1 = 1$$

2 Now let's do the same with 2: $2^3 = 8$. We can show 8 as a cube, too, with sides that are two single-unit cubes long.

The cube is made up of eight small cubes

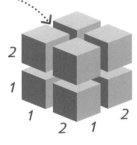

$$2 \times 2 \times 2 = 8$$

3 Next we cube 3: $3^3 = 27$. This cube's sides are three single-unit cubes long.

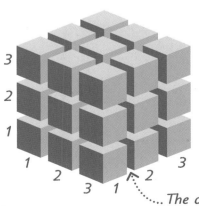

The cube is made up of 27 small cubes

$$3 \times 3 \times 3 = 27$$

4 Next, we calculate that $4^3 = 64$. The new cube has sides that are four single-cube units long.

$$4 \times 4 \times 4 = 64$$

Fractions

A fraction is a part of a whole. We write a fraction as one number over another number. The bottom number tells us how many parts the whole is divided into and the top number says how many parts we have.

What is a fraction?

Fractions are really useful when we need to divide things into equal parts. Let's use this cake to show what we mean when we say something has been divided into quarters.

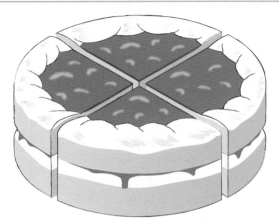

1 The cake has been cut up to make four equal-sized slices, called quarters.

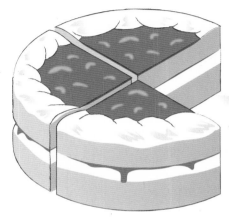

2 Each slice of cake is a quarter of the whole cake. But what does that mean?

Unit fractions

A unit fraction has 1 as its numerator. It is one part of a whole that is divided into equal parts. Let's divide our cake into different unit fractions, up to one-tenth. Can you see that the larger the denominator, the smaller the slice?

A half means "one part out of a possible two parts"

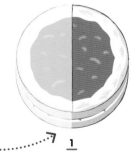

$\frac{1}{2}$
ONE-HALF

$\frac{1}{3}$
ONE-THIRD

$\frac{1}{4}$
ONE-QUARTER

Non-unit fractions

A non-unit fraction has a numerator that is more than one. Fractions can describe parts of a whole, like the cake above, or parts of a group, as with these cakes.

2/5 of the cakes are pink, so 3/5 of them are blue

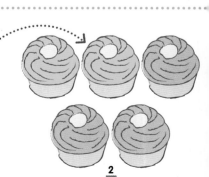

1 There are five cakes. Two of them are pink, so we can say that two-fifths of the cakes are pink.

$\frac{2}{5}$
TWO-FIFTHS ARE PINK

A fraction can be part of one thing, like half a pizza, or part of a group, like half the students in a class.

This shows the robot's ONE slice of cake

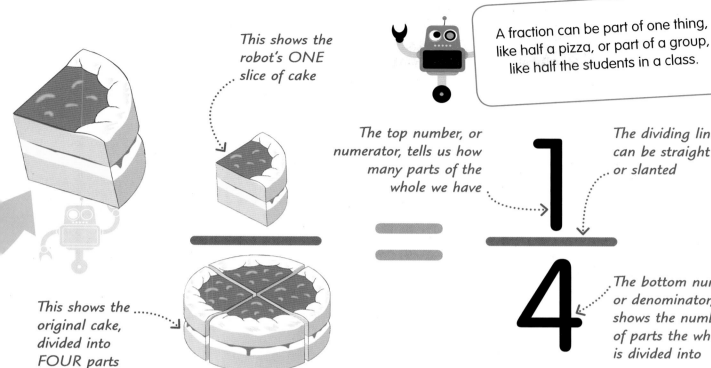

The top number, or numerator, tells us how many parts of the whole we have

The dividing line can be straight or slanted

This shows the original cake, divided into FOUR parts

The bottom number, or denominator, shows the number of parts the whole is divided into

3 It means that each slice is ONE part out of the original cake, which was divided into FOUR parts.

4 We write a fraction as the number of parts we have (the numerator) over the total number of parts (the denominator).

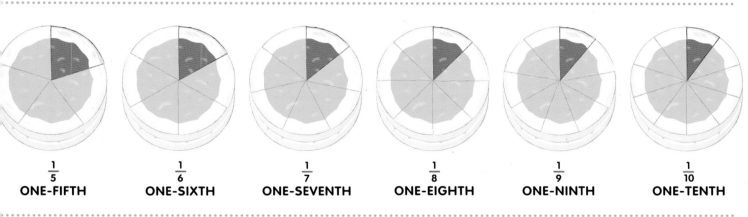

$\frac{1}{5}$	$\frac{1}{6}$	$\frac{1}{7}$	$\frac{1}{8}$	$\frac{1}{9}$	$\frac{1}{10}$
ONE-FIFTH	**ONE-SIXTH**	**ONE-SEVENTH**	**ONE-EIGHTH**	**ONE-NINTH**	**ONE-TENTH**

$5/7$ of the cakes are pink, so $2/7$ of them are blue

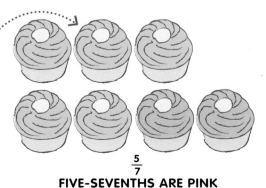

2 This time, there are seven cakes and five are pink. So, five-sevenths of the cakes are pink.

$\frac{5}{7}$
FIVE-SEVENTHS ARE PINK

The cupcake has been divided into thirds

3 Non-unit fractions can be parts of a whole, too. This shows two-thirds of a cake that's been divided into three.

$\frac{2}{3}$
TWO-THIRDS OF A CAKE

Improper fractions and mixed numbers

Fractions aren't always less than a whole. When we want to show that the number of parts is greater than a whole, we can write the result as an improper fraction or mixed number.

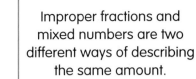

Improper fractions and mixed numbers are two different ways of describing the same amount.

Improper fractions

In an improper fraction, the numerator is larger than the denominator. This tells us that the parts make up more than one whole.

$$\frac{1}{2} \quad \frac{1}{2} \quad \frac{1}{2} \quad \frac{1}{2} \quad \frac{1}{2} \quad = \quad \frac{5}{2}$$

There are five parts

Each part is 1/2 of a whole

1 Look at these five pieces of pizza. We can see that each piece is half of a whole pizza, so we can say that we have five lots of half a pizza.

2 We write this as the fraction 5/2. This means that we have five parts, and each part is one half (1/2) of a whole.

Mixed numbers

A mixed number is a whole number together with a proper fraction. It's another way of writing an improper fraction.

$$= \quad 2\frac{1}{2}$$

Whole number

Proper fraction

1 If we put our pizza halves together, we can make two whole pizzas, with one half left over. So, we can also describe the amount of pizza as "two wholes and one half", or "two and a half".

2 We write it like this: 2½. This mixed number is equal to the improper fraction 5/2:

$$2\frac{1}{2} = \frac{5}{2}$$

Changing an improper fraction to a mixed number

1 What would the improper fraction ¹⁰/₃ be as a mixed number? The fraction tells us that we have 10 lots of one third (¹/₃).

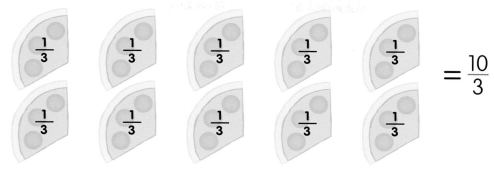

$$= \frac{10}{3}$$

2 If we put the thirds together, we can make three wholes, with one third left over. We can write this as a mixed number: 3 ¹/₃.

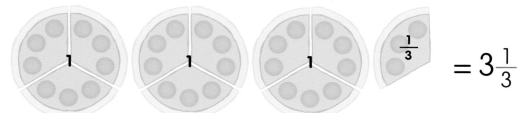

$$= 3\frac{1}{3}$$

3 To make an improper fraction a mixed number, divide the numerator by the denominator. Write down the whole number part of the answer. Then write a fraction in which the numerator is the remainder over the original denominator.

Numerator of the improper fraction ⋯⋯

Denominator of the improper fraction ⋯⋯

$$\frac{10}{3} = 10 \div 3 = 3\frac{1}{3}$$

Changing a mixed number to an improper fraction

1 Let's change 1 ³/₈ into an improper fraction. First, we divide the whole into eighths, because the denominator of the fraction in our mixed number is 8.

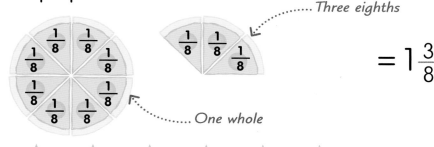

Three eighths

One whole

$$= 1\frac{3}{8}$$

2 If we count the eighths in one whole, then add the three-eighths of our fraction, we have 11 eighths. We write this as the improper fraction ¹¹/₈.

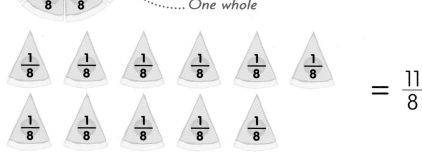

$$= \frac{11}{8}$$

3 To change a mixed number to an improper fraction, we multiply the whole number by the denominator, then add it to the original numerator to make the new numerator.

Denominator ⋯

Whole number ⋯⋯

⋯⋯ Numerator

$$1\frac{3}{8} = \frac{1 \times 8 + 3}{8} = \frac{11}{8}$$

Equivalent fractions

The same fraction can be written in different ways – for example, half a pizza is exactly the same amount as two quarters. We call these equivalent fractions.

1 Look at this table, called a fraction wall. It shows different ways to divide a whole into different unit fractions.

2 Look at the second row, which shows halves, and compare it to the row of fourths, or quarters. We can see that ½ takes up the same amount of the whole as ²/₄.

3 Now we know that ½ and ²/₄ are equal and describe the same fraction of a whole. So, we call ½ and ²/₄ equivalent fractions.

This line helps us to see the amount of the whole taken up by one half (½)

Two quarters takes up the same space as one half

Follow the line down to see which other fractions are the same as one half

To make equivalent fractions, we multiply or divide the numerator and the denominator by the same number.

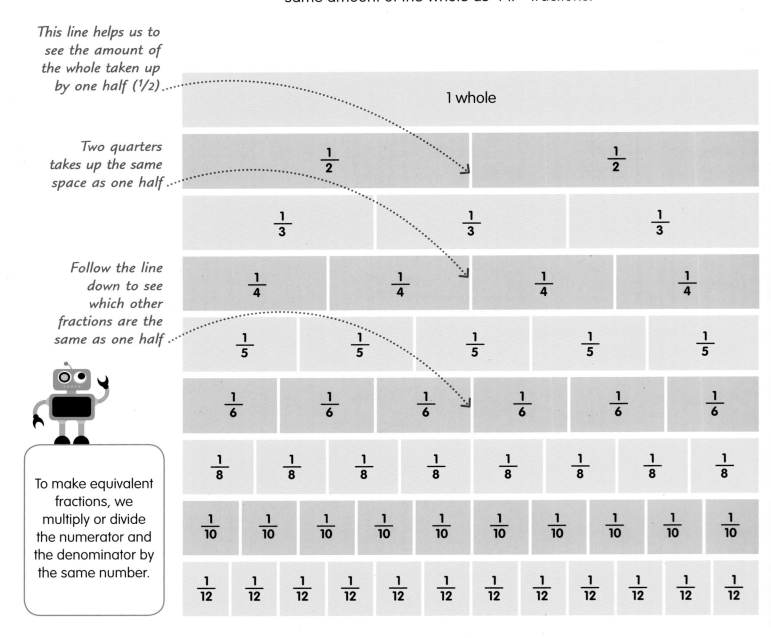

Calculating equivalent fractions

To change a fraction to an equivalent fraction, we multiply or divide the numerator and denominator by a whole number, making sure we use the same whole number both times!

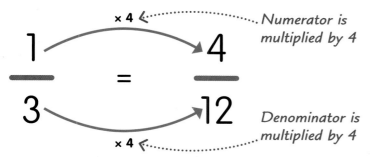

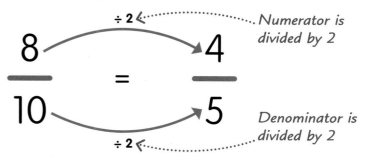

1 Multiplying
We can make $\frac{1}{3}$ into the equivalent fraction $\frac{4}{12}$ by multiplying the numerator and the denominator by 4. Look at the table opposite to check that the two fractions are equivalent.

2 Dividing
We can change $\frac{8}{10}$ into an equivalent fraction by dividing the numerator and the denominator by 2 to make $\frac{4}{5}$. Look at the table on the opposite page to check that $\frac{8}{10}$ and $\frac{4}{5}$ are equivalent.

Using a multiplication grid to find equivalent fractions

We usually use this grid to help us multiply numbers, as on page 106, but it's also a quick and easy way to find equivalent fractions!

1 Look at the top two rows, beginning 1 and 2. Imagine a dividing line between them, making the two rows into fractions, like this:

$$\frac{1}{2} \quad \frac{2}{4} \quad \frac{3}{6} \quad \frac{4}{8} \quad \frac{5}{10} \quad ...$$

2 The first fraction we have is $\frac{1}{2}$. If we read right along the row, we find that all the other fractions, up to $\frac{12}{24}$, are equivalent to $\frac{1}{2}$.

3 This works even for rows that aren't next to each other in the table. So, if we put rows 7 and 11 together, we get a row of fractions that are equivalent to $\frac{7}{11}$:

$$\frac{7}{11} \quad \frac{14}{22} \quad \frac{21}{33} \quad \frac{28}{44} \quad \frac{35}{55} \quad ...$$

×	1	2	3	4	5	6	7	8	9	10	11	12
1	1	2	3	4	5	6	7	8	9	10	11	12
2	2	4	6	8	10	12	14	16	18	20	22	24
3	3	6	9	12	15	18	21	24	27	30	33	36
4	4	8	12	16	20	24	28	32	36	40	44	48
5	5	10	15	20	25	30	35	40	45	50	55	60
6	6	12	18	24	30	36	42	48	54	60	66	72
7	7	14	21	28	35	42	49	56	63	70	77	84
8	8	16	24	32	40	48	56	64	72	80	88	96
9	9	18	27	36	45	54	63	72	81	90	99	108
10	10	20	30	40	50	60	70	80	90	100	110	120
11	11	22	33	44	55	66	77	88	99	110	121	132
12	12	24	36	48	60	72	84	96	108	120	132	144

Simplifying fractions

Simplifying a fraction means reducing the size of the numerator and denominator to make an equivalent fraction that's easier to work with.

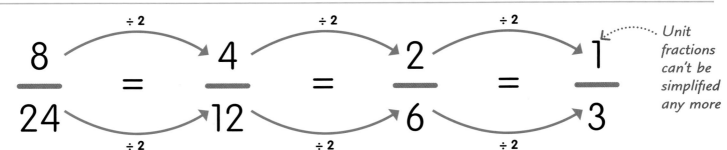

Unit fractions can't be simplified any more

1 Let's look at $^8/_{24}$. If we divide the numerator and denominator by 2, we make an equivalent fraction: $^4/_{12}$.

2 Can we simplify $^4/_{12}$? If we divide both the numerator and denominator by 2 again, we get $^2/_6$.

3 Now we can simplify $^2/_6$ by dividing both parts by 2 again, to get $^1/_3$.

4 The numerator and denominator of $^1/_3$ can't be divided any more, so our fraction is now in its simplest form.

Simplifying fractions using the highest common factor

Instead of going through several stages to simplify a fraction, we can do it by dividing both the numerator and the denominator by their highest common factor (HCF). Remember, we looked at common factors on page 29.

1 Let's simplify the fraction $^{15}/_{21}$. Using the method we learned on page 29, we first list all the factors of the numerator, 15. They are 1, 3, 5, and 15.

2 Now we find the factors of the denominator, 21. They are 1, 3, 7, and 21. The common factors of the numerator and the denominator are 1 and 3, with 3 being the highest common factor.

3 So, if we divide the numerator and the denominator by 3, we get $^5/_7$. We have worked out that $^5/_7$ is the simplest fraction we can make from $^{15}/_{21}$.

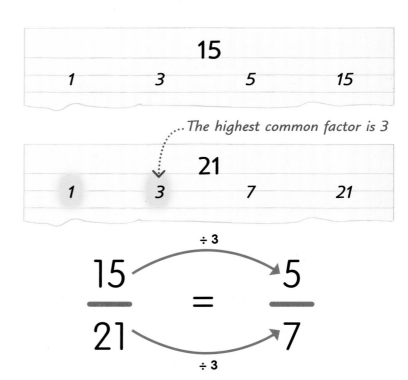

The highest common factor is 3

Finding a fraction of an amount

Sometimes, we need to find out exactly what a fraction of a number or an amount is. Here's how to do it.

To find a fraction of an amount, divide the amount by the denominator, then multiply the answer by the numerator.

1 Look at this herd of 12 cows. How many cows would two-thirds of the herd be?

$\frac{2}{3}$ of 12 = ?

2 First, we find one-third of 12 by dividing it by 3, the denominator of the fraction. The answer is 12 ÷ 3 = 4, so one-third of the herd is four cows.

$\frac{1}{3}$ of 12 = 4

3 We know that one-third of 12 is 4, so to find two-thirds, we multiply 4 by 2. The answer is 4 × 2 = 8, so we know that two-thirds of 12 is 8.

$\frac{2}{3}$ of 12 = 8

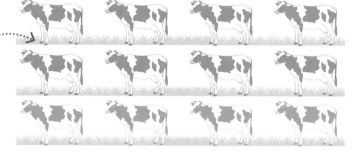

TRY IT OUT

Count your chickens

A farmer has a flock of 24 chickens. If he decided to sell 3/4 of his flock, how many would he take to the market?

Answer on page 319

Comparing fractions with the same denominators

When we need to compare and order fractions, the first thing we do is look at the denominators. If the denominators are the same, all we need to do is put the numerators in order.

1 Look at these fractions. How can we put them in order, from smallest to largest?

2 All the fractions have the same denominator, 8. Remember, the denominator is the number at the bottom of a fraction that tells us how many equal parts a whole has been divided into.

3 Because these denominators are all the same, all we need to do to compare the fractions is look at the numerators.

4 The numerator tells us how many parts of the whole we have. A bigger numerator means more parts. So, let's put the fractions in ascending order (from smallest to largest).

5 If we show these fractions as peas in a pod, it's easy to see which ones are smallest and largest.

When the denominators are the same, we can say that the larger the numerator, the greater the fraction.

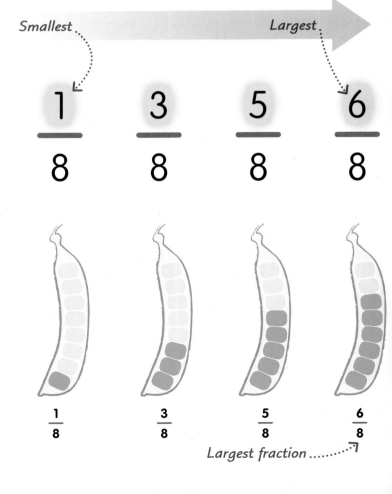

Numerator

$$\frac{5}{8} \quad \frac{3}{8} \quad \frac{1}{8} \quad \frac{6}{8}$$

Denominator

Smallest   Largest

$$\frac{1}{8} \quad \frac{3}{8} \quad \frac{5}{8} \quad \frac{6}{8}$$

$\frac{1}{8} \qquad \frac{3}{8} \qquad \frac{5}{8} \qquad \frac{6}{8}$

Largest fraction

Comparing unit fractions

Unit fractions are fractions where the numerator is 1.
To compare unit fractions, we compare their different
denominators and put them in order.

1 Take a look at these jumbled fractions.
Let's try to put them in ascending order.

2 These fractions all have the same
numerator, 1. Each of these fractions
is just one part of a whole.

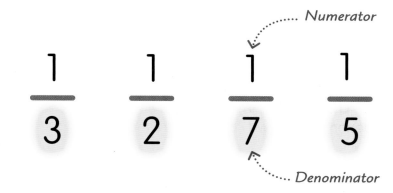

3 We can compare them by looking at the
denominators. A bigger denominator
means the whole is split into more equal parts.

4 The more parts we split the whole into,
the smaller the parts will be. So, the larger
the denominator, the smaller the fraction. Let's
use the denominators to put the fractions in order,
from smallest to largest.

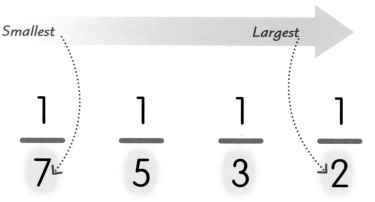

5 If we show these fractions as parts of a whole
carrot, we can see how each portion gets
smaller when the denominator is greater.

When the numerators are
the same, we can say that the
smaller the denominator,
the greater the fraction.

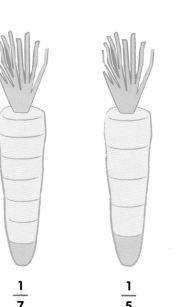
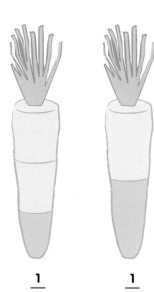

$$\frac{1}{7} \qquad \frac{1}{5} \qquad \frac{1}{3} \qquad \frac{1}{2}$$

Largest fraction

Adding fractions

We add fractions together by adding their numerators, but first we have to make sure they have the same denominator.

To add fractions, we add the numerators and write the total over the common denominator.

Adding fractions that have the same denominator

To add fractions that already have the same denominator, we just add the numerators. So, if we add $2/5$ to $1/5$, we get $3/5$.

Adding two-fifths to one-fifth makes three-fifths

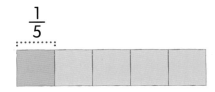

$\frac{1}{5}$ $+$ $\frac{2}{5}$ $=$ $\frac{3}{5}$

Adding fractions that have different denominators

1 Let's try the calculation $2\frac{1}{4} + \frac{1}{6}$. First, we have to change the mixed number into an improper fraction.

$$2\frac{1}{4} + \frac{1}{6} = ?$$

2 We change $2\frac{1}{4}$ to an improper fraction by multiplying 2, the whole number, by 4, the fraction's denominator. Then we add 1, its numerator, to make $9/4$. Now we can write our calculation $9/4 + 1/6$.

$$2\frac{1}{4} = \frac{2 \times 4 + 1}{4} = \frac{9}{4}$$

Both numerator and denominator are multiplied by the same number

3 Next, we give our two fractions the same denominators. Their lowest common denominator is 12, so we make the fractions into twelfths, as we learned on page 51.

$$\frac{9}{4} \xrightarrow{\times 3} = \frac{27}{12} \qquad \frac{1}{6} \xrightarrow{\times 2} = \frac{2}{12}$$

4 goes into 12 three times, so we multiply by 3

6 goes into 12 twice, so we multiply by 2

4 Now we add the numerators of the fractions to make $29/12$. Lastly, we change our answer to a mixed number.

$$\frac{27}{12} + \frac{2}{12} = \frac{29}{12} \quad \text{so} \quad 2\frac{1}{4} + \frac{1}{6} = 2\frac{5}{12}$$

The improper fraction $29/12$ is changed to a mixed number

Subtracting fractions

To subtract fractions, first we check they have the same denominators.
Then we just subtract one numerator from the other.

Subtracting fractions that have the same denominator

To subtract fractions with the same denominator, we simply subtract the numerators. So, if we subtract ¼ from ¾, we get ²⁄₄ , or ½.

Two of the original three-quarters are left

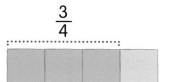

$$\frac{3}{4} \qquad - \qquad \frac{1}{4} \qquad = \qquad \frac{2}{4} \text{ or } \frac{1}{2}$$

Subtracting fractions that have different denominators

1 Let's try the calculation $3\frac{1}{2} - \frac{2}{5}$. As with adding fractions, first we need to change the mixed number and make the fractions' denominators the same.

$$3\frac{1}{2} - \frac{2}{5} = ?$$

2 We change $3\frac{1}{2}$ to an improper fraction by multiplying the whole number by 2, the fraction's denominator, then adding 1, its numerator, to make $\frac{7}{2}$.

$$3\frac{1}{2} = \frac{3 \times 2 + 1}{2} = \frac{7}{2}$$

3 Now we rewrite the fractions so they have the same denominator. The lowest common denominator of $\frac{7}{2}$ and $\frac{2}{5}$ is 10, so we change our two fractions into tenths.

$$\overset{\times 5}{\frac{7}{2}} = \frac{35}{10} \qquad \overset{\times 2}{\frac{2}{5}} = \frac{4}{10}$$

$\times 5$ $\times 2$

2 goes into 10 five times, so the numerator and denominator are multiplied by 5

5 goes into 10 twice, so we multiply by 2

4 We can now subtract one numerator from the other like this: $\frac{35}{10} - \frac{4}{10} = \frac{31}{10}$. We finish by changing $\frac{31}{10}$ back into a mixed number.

$$\frac{35}{10} - \frac{4}{10} = \frac{31}{10} \quad \text{so} \quad 3\frac{1}{2} - \frac{2}{5} = 3\frac{1}{10}$$

Multiplying fractions

Let's look at how to multiply a fraction by a whole number
or by another fraction.

Multiplying by whole numbers and by fractions

What happens when we multiply by a fraction? Let's multiply 4 by a whole number, and by a proper fraction. Remember, a proper fraction is less than 1.

The answer is larger than the original number

$$4 \times 2 = 8$$

1 Multiplying by a whole number
When we multiply 4 by 2, we get 8. This is what we'd expect – that multiplying a number makes it bigger.

The answer is smaller than the original number

$$4 \times \tfrac{1}{2} = 2$$

2 Multiplying by a fraction
Multiplying 4 by ½ makes 2. When we multiply by a proper fraction, the answer is always smaller than the original number.

Multiplying a fraction by a whole number

Let's look at some different calculations to work out what happens when we multiply fractions.

1 Let's try the calculation ½ × 3. This is the same as three groups of one half, so we can add three halves together on a number line to make 1½.

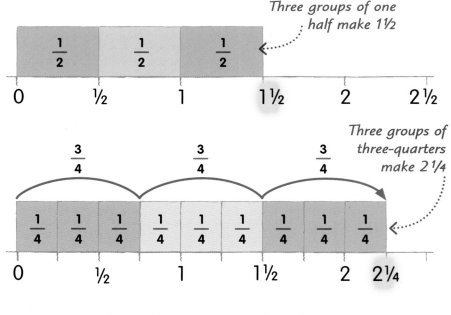

Three groups of one half make 1½

2 Now let's work out ¾ × 3 on a number line. If we add all the quarters in three groups of three-quarters, we get 2¼.

Three groups of three-quarters make 2¼

3 To work out the same calculations without a number line, we simply multiply the whole number by the fraction's numerator, like this.

$$\frac{1}{2} \times 3 = \frac{1 \times 3}{2} = \frac{3}{2} \text{ or } 2\frac{1}{2}$$

$$\frac{3}{4} \times 3 = \frac{3 \times 3}{4} = \frac{9}{4} \text{ or } 2\frac{1}{4}$$

Multiplying fractions with a fraction wall

When we multiply two fractions together, it can be useful to say that the "×" symbol means "of". Let's find out how this works with the help of a fraction wall.

This section is one half of the original quarter

1 whole

$\frac{1}{4}$	$\frac{1}{4}$	$\frac{1}{4}$	$\frac{1}{4}$

1 whole

1 whole

$\frac{1}{8}$	$\frac{1}{8}$	$\frac{1}{8}$	$\frac{1}{8}$
$\frac{1}{8}$	$\frac{1}{8}$	$\frac{1}{8}$	$\frac{1}{8}$

1 For the calculation ½ × ¼, let's say this means "one half of one quarter". First, let's divide a whole into four quarters and shade in one quarter.

$$\frac{1}{2} \times \frac{1}{4} = ?$$

2 Now to find one half of the quarter, we draw a line through the middle of the four quarters. By dividing each quarter in half, we now have eight equal parts.

The calculation ½ × ¼ is the same as saying "a half of a quarter".

3 Let's shade in the top half of our original quarter. This part is one half of a quarter, and also one-eighth of the whole. So we can say that ½ × ¼ = ⅛.

$$\frac{1}{2} \times \frac{1}{4} = \frac{1}{8}$$

How to multiply fractions

Let's look at another way we can multiply fractions, without drawing a fraction wall.

To multiply fractions we multiply the numerators to make a new numerator. Then we multiply the denominators to make a new denominator.

1 Look at this calculation. Can you see that the numerators and the denominators have been multiplied together to make the answer?

$$\frac{1}{2} \times \frac{1}{6} = ?$$

Multiply the numerators together

Multiply the denominators together

$$\frac{1}{2} \times \frac{1}{6} = \frac{1 \times 1}{2 \times 6} = \frac{1}{12}$$

2 Now let's try with two non-unit fractions. The method is exactly the same – just multiply the numerators and the denominators to find the answer.

$$\frac{2}{5} \times \frac{2}{3} = ?$$

Multiply the numerators together

Multiply the denominators together

$$\frac{2}{5} \times \frac{2}{3} = \frac{2 \times 2}{5 \times 3} = \frac{4}{15}$$

Dividing fractions

Dividing a whole number by a proper fraction makes it larger. We can divide fractions using a fraction wall, but there's also a written way to do it.

Dividing by whole numbers and by fractions

Dividing by a fraction gives a number that's larger than the original one

What happens when we divide a whole number by a proper fraction, compared to dividing it by another whole number? Remember, a proper fraction is a fraction that's less than 1.

$$8 \div 2 = 4$$

1 **Dividing by a whole number**
When we divide 8 by 2, the answer is 4. This is what we'd expect – that dividing a number makes it smaller.

$$8 \div \tfrac{1}{2} = 16$$

2 **Dividing by a proper fraction**
When we divide 8 by ½, we are finding how many halves there are in 8. The answer is 16, which is larger than 8.

Dividing a fraction by a whole number

Why does dividing a fraction by a whole number give a smaller fraction? We can use a fraction wall to find out.

$$\tfrac{1}{2} \div 2 = ?$$

When a half is divided into two equal parts, each part is a quarter of the whole

1 We can think of ½ ÷ 2 as "one half shared between two". The fraction wall shows that if we share a half into two equal parts, each new part is one-quarter of the whole.

$$\tfrac{1}{2} \div 2 = \tfrac{1}{4}$$

$$\tfrac{1}{4} \div 3 = ?$$

One-quarter can be divided into three, to make three-twelfths

2 Now let's try ¼ ÷ 3. On the fraction wall, we can see that when one-quarter is divided into three equal parts, each new part is one-twelfth of the whole.

$$\tfrac{1}{4} \div 3 = \tfrac{1}{12}$$

How to divide a fraction by a whole number

There's a simple way to divide a fraction by a whole number – by turning things upside down!

1 Look at these calculations. Can you see a pattern? We can make the denominators of the answers by multiplying the whole numbers and the denominators together. We can use this pattern to divide by fractions without using a fraction wall.

$$\frac{1}{2} \div 8 = \frac{1}{16}$$

$$\frac{1}{3} \div 2 = \frac{1}{6}$$

$$\frac{1}{4} \div 3 = \frac{1}{12}$$

If we multiply the original denominator by the whole number, we get the new denominator

If we multiply 4 and 3 together, we get 12

2 Let's work out ½ ÷ 3. First, we have to make the whole number into a fraction.

$$\frac{1}{2} \div 3 = ?$$

3 To write the number 3 as a fraction, we make 3 the numerator over a denominator of 1, like this.

$$3 = \frac{3}{1}$$

The whole number becomes the numerator

When we write a whole number as a fraction, the denominator is always 1

4 Next, we turn our new fraction upside down and change the division sign into a multiplication sign. So our calculation is now ½ × ⅓.

The ÷ sign changes to a × sign

The denominator becomes the numerator

$$\frac{1}{2} \div \frac{3}{1} = \frac{1}{2} \times \frac{1}{3}$$

The numerator becomes the denominator

5 Now we just have to multiply the two numerators, then the two denominators, to get the answer, ⅙.

$$\frac{1}{2} \div 3 = \frac{1}{2} \times \frac{1}{3} = \frac{1}{6}$$

TRY IT OUT

Division revision

Now it's your turn! Try out your fraction division skills with these tricky teasers.

Answers on page 319

1 ⅙ ÷ 2 = ?   **2** ½ ÷ 5 = ?

3 ⅐ ÷ 3 = ?   **4** ⅔ ÷ 4 = ?

Decimal numbers

Decimal numbers are made up of whole numbers and fractions of numbers.
A dot, called a decimal point, separates the two parts of a decimal number.

1 Decimals are really useful when we want to make accurate measurements, such as recording the runners' times in this race.

2 On the scoreboard, the digits to the left of the decimal point show whole seconds. The digits to the right show parts, or fractions, of a second.

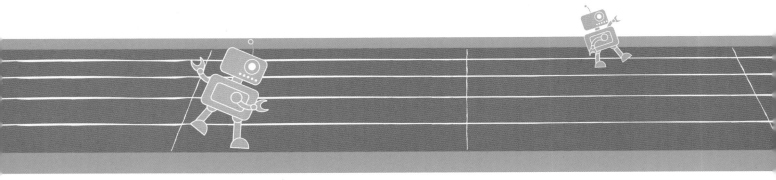

Decimals are fractions, too!

The digits after the point in a decimal number are just another way of showing fractions, or numbers less than one. Let's find out how they work.

1 Tenths
If we put $2\frac{7}{10}$ into place-value columns, the whole number 2 goes in the ones column and the 7 in the tenths column to stand for $\frac{7}{10}$. So we can also write $2\frac{7}{10}$ as 2.7.

$$2\frac{7}{10} = \begin{array}{c|c|c} O & \frac{1}{10} & \\ \hline 2 & . & 7 \end{array}$$

The 7 in the tenths column stands for $\frac{7}{10}$

2 Hundredths
Now let's do the same with $2\frac{72}{100}$. When we put all the digits into their place-value columns, we can see that $2\frac{72}{100}$ is the same as 2.72.

$$2\frac{72}{100} = \begin{array}{c|c|c|c} O & \frac{1}{10} & \frac{1}{100} \\ \hline 2 & . & 7 & 2 \end{array}$$

This 2 stands for $\frac{2}{100}$

3 Thousandths
Finally, when we put $2\frac{721}{1000}$ into place-value columns, we see that $2\frac{721}{1000}$ is the same as 2.721.

This 1 stands for $\frac{1}{1000}$

$$2\frac{721}{1000} = \begin{array}{c|c|c|c|c} O & \frac{1}{10} & \frac{1}{100} & \frac{1}{1000} \\ \hline 2 & . & 7 & 2 & 1 \end{array}$$

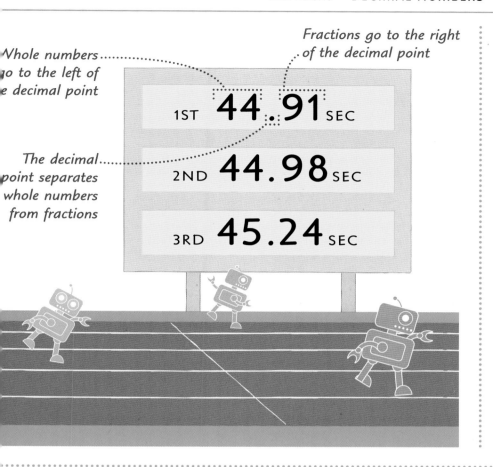

Whole numbers go to the left of the decimal point

Fractions go to the right of the decimal point

The decimal point separates whole numbers from fractions

1ST **44.91** SEC

2ND **44.98** SEC

3RD **45.24** SEC

Fraction converter

Here is a table of some of the most common fractions and their equivalent decimal fractions.

Fraction	Decimal
$\frac{1}{1000}$	0.001
$\frac{1}{100}$	0.01
$\frac{1}{10}$	0.1
$\frac{1}{5}$	0.2
$\frac{1}{4}$	0.25
$\frac{1}{3}$	0.33
$\frac{1}{2}$	0.5
$\frac{3}{4}$	0.75

Rewriting fractions as decimals

To rewrite a fraction as a decimal, we first turn it into an equivalent fraction in tenths, hundredths, or thousandths. We do this by finding a number we can multiply by the fraction's denominator to make it 10, 100, or 1000.

1 **½ is the same as 0.5**
We can change ½ into ⁵⁄₁₀ by multiplying the numerator and denominator by 5. When we put ⁵⁄₁₀ into place-value columns, we get the decimal fraction 0.5.

The numerator is multiplied by 5

The 5 in the tenths column means "five-tenths"

$$\frac{1}{2} = \frac{5}{10}$$

× 5

× 5

O	$\frac{1}{10}$
0 .	5

The denominator is multiplied by 5

2 **¼ is the same as 0.25**
We can change ¼ into ²⁵⁄₁₀₀ by multiplying it by 25. When we put the new fraction into place-value columns, we see that ²⁵⁄₁₀₀ is 0.25.

$$\frac{1}{4} = \frac{25}{100}$$

× 25

× 25

O	$\frac{1}{10}$	$\frac{1}{100}$
0 .	2	5

²⁵⁄₁₀₀ is the same as 0.25

Comparing and ordering decimals

When we compare decimals, we look at the digits with the highest place values first.

When we compare or order decimals, we use what we know about place value, just as we do when we compare whole numbers.

Comparing decimals

When we compare decimals, we compare the digits with the highest place value first to decide which number is larger.

O	$\frac{1}{10}$	$\frac{1}{100}$
0 .	1	
0 .	0	1

The placeholder, zero, tells us there are no tenths

1 **0.1 is greater than 0.01**
The digits in the ones column are the same, so we compare the digits in the tenths column to find that 0.1 is the greater number.

O	$\frac{1}{10}$	$\frac{1}{100}$
2 .	6	1
2 .	6	5

5 is greater than 1 so 2.65 is the larger number

2 **2.65 is greater than 2.61**
This time we have to compare the hundredths columns to find that the greater number of the two is 2.65.

Ordering decimals

On page 22, we found out how to put whole numbers in order. Ordering decimals works in just the same way!

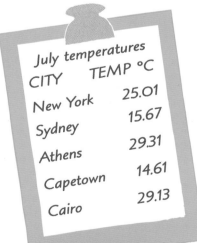

July temperatures

CITY	TEMP °C
New York	25.01
Sydney	15.67
Athens	29.31
Capetown	14.61
Cairo	29.13

We compare the digits in order, starting with the most significant

	T	O	$\frac{1}{10}$	$\frac{1}{100}$
Athens	2	9 .	3	1
Cairo	2	9 .	1	3
New York	2	5 .	0	1
Sydney	1	5 .	6	7
Capetown	1	4 .	6	1

1 Let's help sun-loving Kloog choose a holiday hotspot by putting his list of cities in order, with the highest temperature first. As with whole numbers, we order decimal numbers by comparing their significant digits.

2 To find the greatest number, we compare each number's most significant digit. If they are the same, we look at the second digits, and then, if necessary, the third, and so on. We carry on comparing until we have ordered the numbers.

Rounding decimals

We round decimals in the same way as we round whole numbers (see pages 26-27). The easiest way to see how it works is by looking at a number line.

> The rounding rule for decimals and whole numbers is the same: digits less than 5 are rounded down, and digits of 5 or more are rounded up.

1 Rounding to one
This means that we round a decimal to the nearest whole number. So 1.3 is rounded down to 1 and 1.7 is rounded up to 2.

1.3 is nearer to 1 than 2, so we round it down

1.7 is nearer to 2 than 1, so it's rounded up ···

1   1.3   1.7   2

2 Rounding to tenths
This means that we round a decimal number to one digit after the decimal point. So 1.12 rounds down to 1.1, and 1.15 rounds up to 1.2.

Digits 4 or less are rounded down

Digits 5 or more are rounded up ····

1.1   1.11   1.12   1.13   1.14   1.15   1.16   1.17   1.18   1.19   1.2

3 Rounding to hundredths
Rounding to hundredths gives us a number with two digits after the decimal point. So 1.114 rounds down to 1.11 and 1.116 rounds up to 1.12.

1.114 rounds down to 1.11

1.116 rounds up to 1.12 ····

1.11   1.111   1.112   1.113   1.114   1.115   1.116   1.117   1.118   1.119   1.12

TRY IT OUT

Decimals workout

Here's a list of the racers' times for the slalom skiing race on Megabyte Mountain. Can you round all their times to hundredths, so there are two digits after the decimal point? Who had the fastest time?

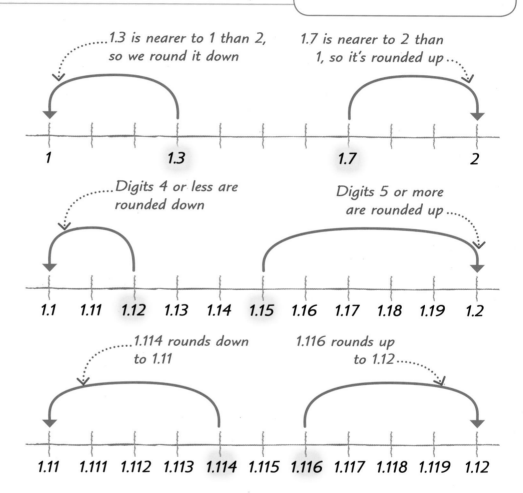

FINISH

TWERG	17.239 SEC
BLOOP	16.560 SEC
GLOOK	17.211 SEC
KWONK	16.129 SEC
ZARG	16.011 SEC

Answer on page 319

Adding decimals

We add decimals in the same way as we add whole numbers –
turn to page 87 to find out the written way to add decimals.

1 Let's add 4.5 and 7.7. To help us see how adding decimals works, we'll show the calculation using counting cubes.

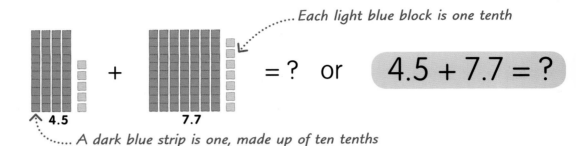

... Each light blue block is one tenth

= ?   or $4.5 + 7.7 = ?$

4.5   **7.7**

... A dark blue strip is one, made up of ten tenths

2 First, let's add the tenths from the two numbers: 0.5 + 0.7. This gives us $^{12}/_{10}$, or 1.2.

... We exchange ten tenth cubes for a ones strip

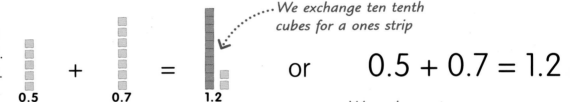

0.5   **0.7**   **1.2**

or $0.5 + 0.7 = 1.2$

3 Now let's add the two whole numbers, 4 and 7, together to make 11.

...We exchange ten one strips for one ten block

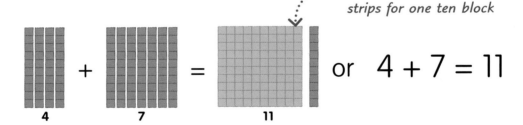

4   **7**   **11**

or $4 + 7 = 11$

4 Now we can add our two answers, 1.2 and 11, together to get the final answer, 12.2.

We have one ten, two ones, and two tenths

11   **1.2**   **12.2**

or $11 + 1.2 = 12.2$

5 We have found that 4.5 + 7.7 = 12.2. When we write the calculation, it looks like this – go to page 87 for more about adding decimals in this way.

T	O	$\frac{1}{10}$
	1	
	4 .	5
+	7 .	7
1	2 .	2

so $4.5 + 7.7 = 12.2$

Subtracting decimals

When we subtract decimal numbers, we use the same method as we do for whole numbers.

1 Let's try the calculation 8.2 – 4.7. We'll use the counting cubes to help us see what happens.

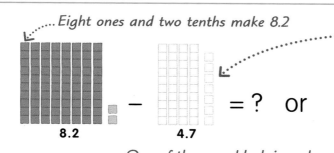

Eight ones and two tenths make 8.2

We are taking away four ones and seven tenths from our original number, 8.2

$$8.2 - 4.7 = ?$$

2 First, let's subtract 0.7, the decimal part of 4.7, from 8.2. We exchange a ones strip for ten tenth cubes so we can take away seven tenths. The answer is 7.5.

One of the ones block is exchanged for ten tenths

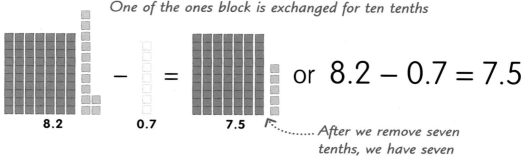

or $8.2 - 0.7 = 7.5$

After we remove seven tenths, we have seven ones and five tenths left

3 Now let's subtract 4, the whole number, from 7.5. When we remove four of the ones strips, we have 3.5 left.

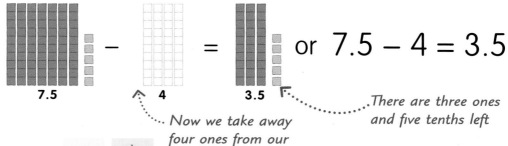

or $7.5 - 4 = 3.5$

There are three ones and five tenths left

Now we take away four ones from our number, 7.5

4 So 8.2 – 4.7 = 3.5. We can write the calculation in columns, like this. Find out more about column subtraction on pages 96-97.

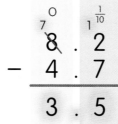

so $8.2 - 4.7 = 3.5$

TRY IT OUT

Over to you!

Find out how much you've learned by trying out these calculations.

Answers on page 319

1 $0.2 + 3.9 = ?$ **2** $45.6 - 21.2 = ?$

3 $10.2 + 21.6 = ?$ **4** $96.7 - 75.8 = ?$

Percentages

Per cent means "per hundred". It shows an amount as part of 100. So, 25 per cent means 25 out of 100. We use the symbol "%" to represent a percentage.

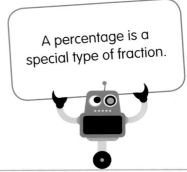

A percentage is a special type of fraction.

Parts of 100

A percentage is a useful way of comparing quantities. For example, in this block of 100 robots, the robots are divided into different colour groups according to the percentage they represent.

1 **1%**
There is only one green robot out of a total of 100. We can write this as 1%. This is the same as $\frac{1}{100}$ or 0.01.

2 **10%**
In the yellow group, there are 10 robots out of 100. We can write this as 10%. This is the same as $\frac{1}{10}$ or 0.1.

3 **50%**
There are 50 robots out of 100 in the red group. We can write this as 50%. This is the same as $\frac{1}{2}$ or 0.5.

4 **100%**
All the robots added together – green, grey, yellow, and red – represent 100%. This is the same as $\frac{100}{100}$ or 1.

TRY IT OUT

Shaded parts

These grids have 100 squares. What percentage is shaded dark purple in each grid?

Answers on page 319

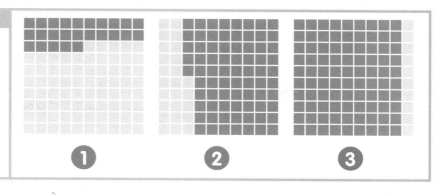

①　②　③

Percentages, decimals, and fractions

We can use a percentage, decimal, and fraction to write the same number. Some of the most common percentages are listed below, together with the decimal and fraction equivalents. You can find out more on pages 74-75.

PERCENTAGE	DECIMAL	FRACTION
1%	0.01	$\frac{1}{100}$
5%	0.05	$\frac{5}{100}$
10%	0.1	$\frac{1}{10}$
20%	0.2	$\frac{1}{5}$
25%	0.25	$\frac{1}{4}$
50%	0.5	$\frac{1}{2}$
75%	0.75	$\frac{3}{4}$
100%	1	$\frac{100}{100}$

Calculating percentages

We can find a percentage of any total amount, not just 100.
The total can be a number or a quantity, such as the area of
a shape. Sometimes we might also want to write one number
as a percentage of another number.

Finding a percentage of a shape

On pages 64-65, we looked at percentages using a square grid divided into 100 parts. But what if a shape has 10 parts or even 20?

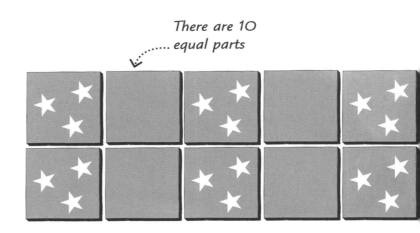

There are 10 equal parts

1 Take a look at this example. There are 10 tiles altogether. What percentage of the tiles have a pattern?

2 The whole amount of any shape is 100%. To find the percentage represented by one part, we divide 100 by the number of parts (10). This gives us 10, so one tile equals 10%.

$$100 \div 10 = 10$$

Each tile is worth 10%

The total number of tiles

3 We multiply the result (10) by the number of patterned tiles (6). This gives us the answer 60. So, 60% of the tiles have a pattern.

$$10 \times 6 = 60$$

60% have a pattern

TRY IT OUT

Working it out

Here are several shapes. What percentage of each shape has been shaded a dark colour?

Answers on page 319

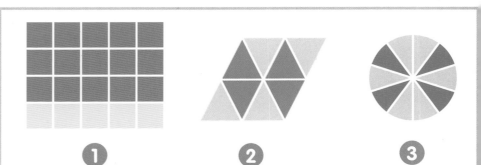

① ② ③

Finding a percentage of a number

We can also use percentages to divide a number into parts. There's more than one way to do this, but one method is to start by finding 1%.

A percentage is just another way of writing a fraction.

1 Let's find 30% of 300.

$$30\% \text{ of } 300 = ?$$

2 First, we need to find 1% of 300, so we divide the 300 by 100.

$$300 \div 100 = 3$$

Divide the total amount by 100

3 Next, we multiply the answer by the percentage we need to find.

$$3 \times 30 = 90$$

4 This gives us the answer: 30% of 300 is 90.

$$30\% \text{ of } 300 = 90$$

The 10% method

In the example above, we began by finding 1% of the total. Sometimes, we can get to the answer more quickly by first finding 10%. This is called the 10% method.

1 In this example, we need to work out 65% of £350.

$$65\% \text{ of } £350 = ?$$

2 We need to find 10% of £350, so we divide the amount by 10. This gives us 35.

$$350 \div 10 = 35$$

3 We know that 10% is 35, so 60% will be 6 groups of 35.

$$6 \times 35 = 210$$

4 We've found 60% of 350. Now we just need another 5% to get 65%. To work out 5%, we simply halve the 10% amount.

$$35 \div 2 = 17.50$$

5 Now add 60% and 5% to find 65%. So, 65% of £350 is £227.50.

$$210 + 17.50 = £227.50$$

TRY IT OUT

10% challenge

Time yourself and see how quickly you can work out the following percentages:

1 10% of 200

2 10% of 550

3 10% of 800

Answers on page 319

Percentage changes

We can use a percentage to describe the size of a change in a number or a measurement. We might also want to work out how much an actual value has increased or decreased when we already know how much it has changed as a percentage.

Calculating a percentage increase

1 This snack bar used to weigh 60g but it's now 12g heavier. What is the percentage increase in the bar's weight?

$$12g = ? \% \text{ of } 60g$$

2 First, we divide the increase in weight by the original weight. This is 12 ÷ 60. The answer is 0.2.

The amount of the change ⌐⌐⌐⌐⌐⌐⌐ ⌄   *The original amount* ⌐⌐⌐ ⌄

$$12 \div 60 = 0.2$$

3 Then we multiply the result by 100. So, we need to work out 0.2 × 100. The answer is 20.

$$0.2 \times 100 = 20$$

4 This means the new bar weighs 20% more than it did before.

$$12g = 20\% \text{ of } 60g$$

Calculating a percentage decrease

1 Here's another snack bar. It used to contain 8g of sugar. To make it healthier, it's now made with 2g less sugar. Let's work out how much the amount of sugar has decreased as a percentage.

$$2g = ? \% \text{ of } 8g$$

2 The first step is to divide the decrease in the amount of sugar by the original amount. This is 2 ÷ 8. The answer is 0.25.

Divide the size of the change by the original amount ⌐⌐⌐⌐⌐ ⌄

$$2 \div 8 = 0.25$$

3 To turn this result into a percentage, we just multiply 0.25 by 100, giving us the answer 25.

$$0.25 \times 100 = 25$$

4 This means the bar now has 25% less sugar.

$$2g = 25\% \text{ of } 8g$$

Turning a percentage increase into an amount

1 One year ago, this bike cost £200. Since then, its price has gone up by 5%. How much more does it cost now?

$$5\% \text{ of } £200 = ?$$

2 First, we need to find 1% of 200. All we need to do is divide 200 by 100. Remember, we looked at dividing by 100 on page 136. The answer is 2.

The original price ⌐ $200 \div 100 = 2$

3 We want to find 5%, so we multiply the value of 1% by 5. This is 2 × 5, and the answer is 10.

1% of the original price ⌐ $2 \times 5 = 10$

4 This means the bike now costs £10 more than it did a year ago.

$$5\% \text{ of } £200 = £10$$

Turning a percentage decrease into an amount

1 Now take a look at this bike. It used to cost £250, but its price has been cut by 30%. If we buy the bike now, how much money will we save?

$$30\% \text{ of } £250 = ?$$

2 Just as in our example with the other bike, the first step is to work out 1% of the original price. This is 250 ÷ 100. The answer is 2.5.

$250 \div 100 = 2.5$ ⌐ *1% of 250*

3 Now we know what 1% is, we can find 30% like this: 2.5 × 30 = 75

$$2.5 \times 30 = 75$$

4 This means the price of the bike has dropped by £75.

$$30\% \text{ of } £250 = £75$$

TRY IT OUT

Percentage values

In a sale, these items have been reduced in price. Can you work out the new prices? To work out the new price, calculate the decrease in price and subtract it from the original price.

Answers on page 319

1 A coat priced £200 has been reduced by 50%.

2 These trainers were £50 but have been reduced by 30%.

3 This T-shirt has been reduced by 10%. It was £15.

Ratio

Ratio is the word we use when we compare two numbers or amounts, to show how much bigger or smaller one is than the other.

> Ratio tells us how much we have of one amount compared to another amount.

1 Let's look at these seven ice cream cones. Three are strawberry and four are chocolate, so we say that the ratio of strawberry to chocolate cones is 3 to 4.

Three strawberry cones ....   *Four chocolate cones* ....

2 The symbol for the ratio between two amounts is two dots on top of each other, so we write the ratio of strawberry to chocolate cones as 3:4.

RATIO OF STRAWBERRY TO CHOCOLATE CONES IS   **3 : 4**

Simplifying ratios

As with fractions, we always simplify ratios when we can. We do this by dividing both numbers in the ratio by the same number.

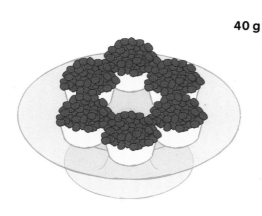

40 g puffed rice cereal

50 g chocolate

40 : 50   **=**   **4 : 5**

Simplify the ratio by dividing both numbers by 10

40:50

÷ 10   ÷ 10

4:5

1 In this recipe, 40 g of puffed rice cereal, plus 50 g of melted chocolate, makes six mini treats.

2 For every 40 g of cereal we use, we need 50 g of chocolate. So the ratio of cereal to chocolate in the recipe is 40:50.

3 To simplify the ratio, we divide both numbers by 10 to make a ratio of cereal to chocolate of 4:5.

Proportion

Proportion is another way of comparing. Instead of comparing one amount with another, as with ratio, proportion is comparing a part of a whole with the whole amount.

Proportion tells us how much we have of something compared to the whole amount.

Proportion as a fraction

We often write proportion as a fraction. Here are 10 cats. What fraction of them is ginger?

1 Four out of the 10 cats are ginger. So, ginger cats make up four-tenths ($^4/_{10}$) of the whole amount.

2 We simplify fractions if we can, so we divide the numerator and denominator of $^4/_{10}$ by 2 to make it $^2/_5$.

We simplify the fraction by dividing both numbers by 2

$$\frac{4}{10} = \frac{2}{5}$$

÷ 2

÷ 2

Four of the 10 cats are ginger

3 So, the proportion of ginger in the whole group, written as a fraction, is $^2/_5$.

PROPORTION OF GINGER CATS = $\dfrac{2}{5}$

Proportion as a percentage

Percentages are another way of writing fractions, so a proportion can be expressed as a percentage, too. What percentage of the cats is grey?

One of the 10 cats is grey

1 There is one grey cat out of 10, so the proportion as a fraction is $^1/_{10}$.

2 To change $^1/_{10}$ into a percentage, we rewrite it as equivalent hundredths, so $^1/_{10}$ becomes $^{10}/_{100}$.

We make an equivalent fraction by multiplying both numbers by 10

× 10

$$\frac{1}{10} = \frac{10}{100}$$

× 10

3 We know that "ten out of one hundred" is the same as 10%, so the percentage of grey cats in the group is 10%.

PROPORTION OF GREY CATS = 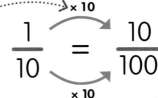 **10%**

Scaling

Scaling is making something larger or smaller while keeping everything in the same proportion – which means making all the parts larger or smaller by the same amount.

We can use scaling to change numbers, amounts, or the sizes of objects or shapes.

Scaling down

A photograph, like this robot selfie, is a perfect example of scaling down.

1 In the photo, the robot looks the same, but smaller. Every part of him has been reduced in size by the same amount.

2 The robot is 75 cm tall in real life. In the photo, he is 15 cm tall. So, he is five times smaller in the photo.

3 The robot's body is 40 cm wide. In the photo, it's 8 cm wide, which is five times smaller than in real life.

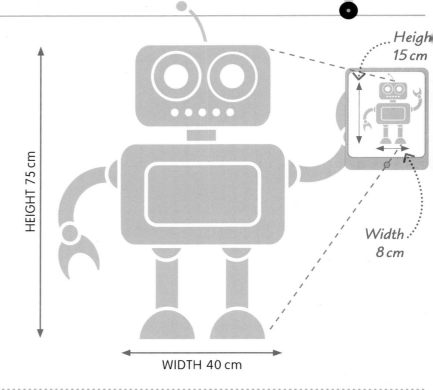

HEIGHT 75 cm

WIDTH 40 cm

Height 15 cm

Width 8 cm

Scaling up

Scaling up is making every part of a thing larger. We can scale up amounts as well as objects and measurements.

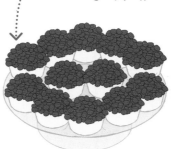

12 treats are made with 100 g of chocolate and 80 g of puffed rice

Chocolate

50 g

? g

Puffed rice

40 g

? g

MAKES 6 CAKES

MAKES 12 CAKES

Multiply both amounts by two

$$50\,g \times 2 = 100\,g$$

$$40\,g \times 2 = 80\,g$$

1 On page 70, we saw a recipe for six chocolate treats. To make 12, we'll need more ingredients. But how much more of each?

2 We know that 12 is 2 times 6. So, if we multiply both ingredients by two, we can make twice as many treats.

3 So, to scale up a recipe, we need to multiply all the ingredients by the same amount.

Scale on maps

Scaling is useful for drawing maps. We couldn't use a life-size map – it would be too big to carry around! We write a map scale as a ratio, which tells us how many units of distance in real life are equal to one unit on the map.

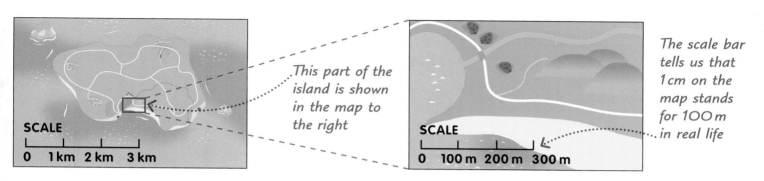

This part of the island is shown in the map to the right

The scale bar tells us that 1 cm on the map stands for 100 m in real life

1 **1 cm : 1 km**
On this map, 1 cm represents 1 km in real life. We can see the whole island, but not in much detail.

2 **1 cm : 100 m**
This time, 1 cm on the map stands for 100 m. We can see lots of detail, but only on a very small part of the island.

Scale factors

A scale factor is the number we multiply or divide by when we scale up or down.

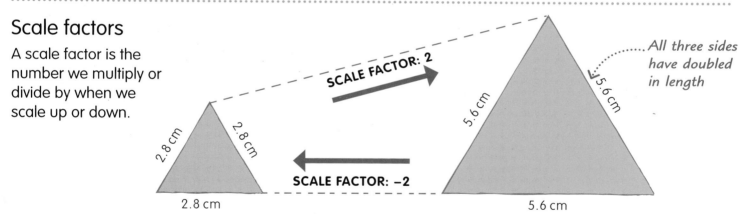

SCALE FACTOR: 2

SCALE FACTOR: –2

2.8 cm

2.8 cm

2.8 cm

5.6 cm

5.6 cm

5.6 cm

All three sides have doubled in length

1 If we scale something by a factor of 2, we make it two times larger. So, this triangle with sides of 2.8 cm becomes a triangle with sides of 5.6 cm.

2 If we scaled the triangle back to its original size, we would say is was scaled by a negative factor of –2.

TRY IT OUT

How tall is a T. rex?

This scale model of a T. rex has a scale factor of 40. If the model's height is 14 cm and its length is 30 cm, can you work out the height and length of the real dinosaur?

Answers on page 319

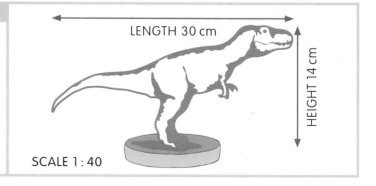

LENGTH 30 cm

HEIGHT 14 cm

SCALE 1 : 40

Different ways to describe fractions

Decimals and percentages are just different ways of describing fractions. Ratio and proportion can be written as fractions, too.

Fractions, decimals, and percentages are all linked, and we can express one as any of the others.

Proportion as a fraction, a decimal, or a percentage

12 out of 20 roses are pink

Look at these 20 roses. There are 12 pink and 8 red roses. Let's describe the proportion of pink roses as a fraction, a decimal, and a percentage.

1 As a fraction
There are 12 pink roses out of a total of 20 roses. So, the proportion of pink roses is $^{12}/_{20}$ or, if we simplify it, $^{3}/_{5}$.

2 As a decimal
If we rewrite $^{3}/_{5}$ as equivalent tenths, we get $^{6}/_{10}$, which is the same as 0.6. So, 0.6 of the group consists of pink roses.

3 As a percentage
If we rewrite $^{6}/_{10}$ as hundredths, we get $^{60}/_{100}$, which can also be written as 60%. So, 60% of the roses are pink.

PROPORTION OF PINK ROSES

$$\frac{3}{5} \quad = \quad 0.6 \quad = \quad 60\%$$

Ratio and fractions

On page 70, we learned how to write ratios using two dots between the numbers. But we can write ratios as fractions, too.

1 Now we have three roses and 12 daisies. We write the ratio of roses to daisies as 3:12, then simplify it to 1:4.

2 We can also write this ratio as $^{3}/_{12}$ or $^{1}/_{4}$, which means that the number of roses is a quarter of the number of daisies.

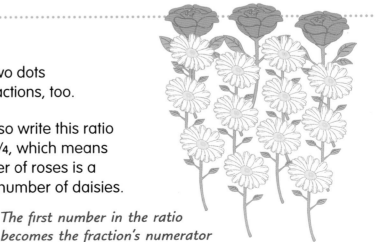

RATIO OF ROSES TO DAISIES

The first number in the ratio becomes the fraction's numerator

$$3:12 \text{ or } 1:4 \quad = \quad \frac{3}{12} \text{ or } \frac{1}{4}$$

The second number in the ratio becomes the fraction's denominato

Common fractions, decimals, and percentages

This table shows the different ways we can show or write the same fraction.

Part of a whole	Part of a group	Fraction in words	Fraction in numbers	Decimal	Percentage
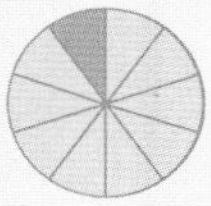	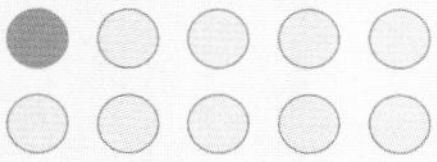	ONE-TENTH	$\frac{1}{10}$	0.1	10%
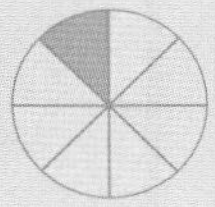	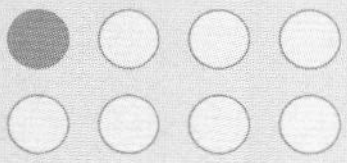	ONE-EIGHTH	$\frac{1}{8}$	0.125	12.5%
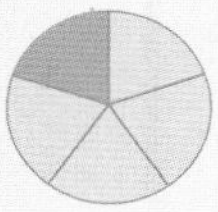		ONE-FIFTH	$\frac{1}{5}$	0.2	20%
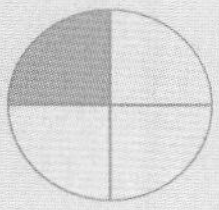		ONE-QUARTER	$\frac{1}{4}$	0.25	25%
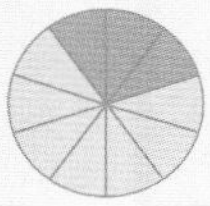	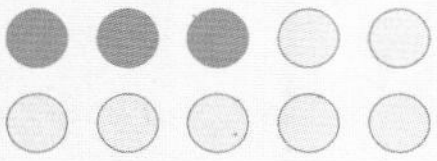	THREE-TENTHS	$\frac{3}{10}$	0.3	30%
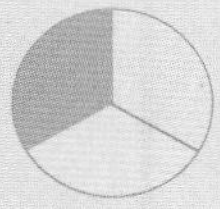		ONE-THIRD	$\frac{1}{3}$	0.33	33%
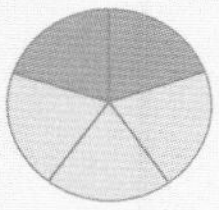		TWO-FIFTHS	$\frac{2}{5}$	0.4	40%
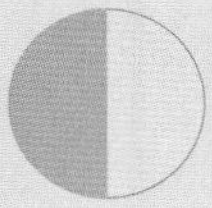		ONE-HALF	$\frac{1}{2}$	0.5	50%
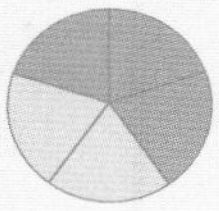		THREE-FIFTHS	$\frac{3}{5}$	0.6	60%
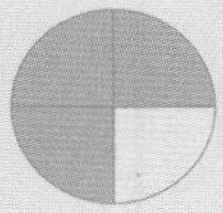		THREE-QUARTERS	$\frac{3}{4}$	0.75	75%

TRY IT OUT

How much do you know?

Try these baffling brainteasers and see if you can get 100% right!

Answers on page 319

1. Write 0.35 as a fraction. Don't forget to simplify it.

2. Write $\frac{3}{100}$ as a percentage, then as a decimal.

3. Write the ratio 4:6 as a fraction. Now simplify it.

We calculate to solve problems in maths.
We can add, subtract, multiply, and divide in
our heads or by writing numbers down on
paper. By learning some useful strategies,
we can work with numbers of any size.
By remembering a few simple rules, we can
also solve calculations in several stages.

Addition

When we bring two or more quantities together to make a larger quantity, it's called addition or adding. There are two ways to think about how addition works.

It doesn't matter which way you add numbers together. The answer will be the same.

What is addition?

Look at these oranges. When we combine 6 oranges and 3 oranges, there are 9 oranges altogether. We can say we have added 6 oranges and 3 oranges, which equals 9 oranges.

This symbol means add or plus

Combining 6 oranges and 3 oranges gives us 9 oranges ...

This symbol means equivalent to, or equals

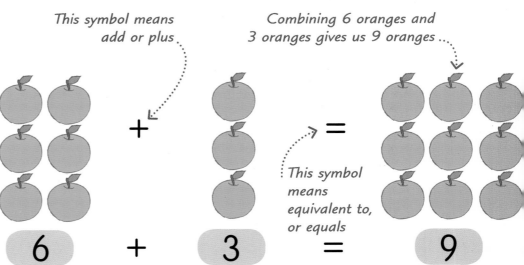

6 + 3 = 9

Adding works in any order

It doesn't matter which way we add amounts. The total will be the same. We say that addition is commutative.

1 Look at this calculation. It says that if we add 2 to 5, we get 7.

5 + 2 = 7

2 Now let's switch the numbers around on the left-hand side of the equals sign. It doesn't matter which order we add numbers, the total will be the same.

2 + 5 = 7

REAL WORLD MATHS

The ancient calculator

The earliest type of calculator was the abacus, used in ancient Egypt, ancient Greece, and other places around the world. The abacus helped people calculate amounts, with beads on different rows used to represent different numbers, like ones, tens, and hundreds.

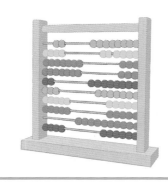

Adding as counting all

We can think of addition as combining two or more amounts into a single amount and then counting them. This way of adding is called counting all.

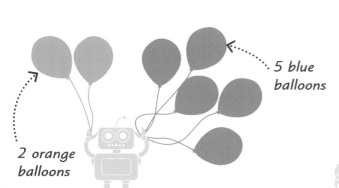

2 orange balloons

5 blue balloons

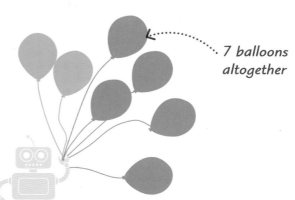

7 balloons altogether

1 Look at these balloons. The robot has 2 balloons in one hand and 5 in the other.

$$2 + 5 = ?$$

2 Now the robot has combined, or added, the balloons by putting them all together in one hand. We can work out the total simply by counting them all. There are 7.

3 So, $2 + 5 = 7$

$$2 + 5 = 7$$

Adding as counting on

There is another way to think about addition. To add one number to another, we can simply count on from the larger number in a series of steps that's equal to the smaller number. This is called counting on.

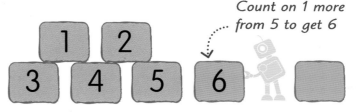

Count on 1 more from 5 to get 6

2 First, he counts on by adding the first red box to get 6.

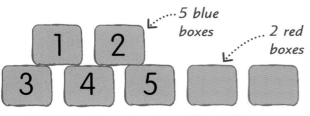

5 blue boxes

2 red boxes

1 This time the robot is adding 5 blue boxes and 2 red boxes. He can do this by counting on from 5.

$$5 + 2 = ?$$

Count on 1 more from 6 to get 7

3 Then he counts on again by adding the second red box to get 7.

$$5 + 2 = 7$$

Adding with a number line

Doing calculations in your head can be tricky. We can use a number line to help us with calculations, including addition. It is most useful for calculations with numbers up to 20.

You can use number lines to work out both addition and subtraction calculations.

1 Let's use a number line to find out the answer when we add 4 and 3.

$$4 + 3 = ?$$

The line doesn't have to be neat — it's just to help you count

2 First, we draw a line and mark it with numbers from 0 to 10.

| O | 1 | 2 | 3 | 4 | 5 | 6 | 7 | 8 | 9 | 10 |

Start counting at 4

3 This calculation starts with the number 4, so first find 4 on the number line.

| O | 1 | 2 | 3 | 4 | 5 | 6 | 7 | 8 | 9 | 10 |

$$1 + 1 + 1 = 3$$

Stop counting at 7

4 We need to add 3 to 4, so next jump 3 places to the right. This takes us to 7.

| O | 1 | 2 | 3 | 4 | 5 | 6 | 7 | 8 | 9 | 10 |

5 So, 4 + 3 = 7

$$4 + 3 = 7$$

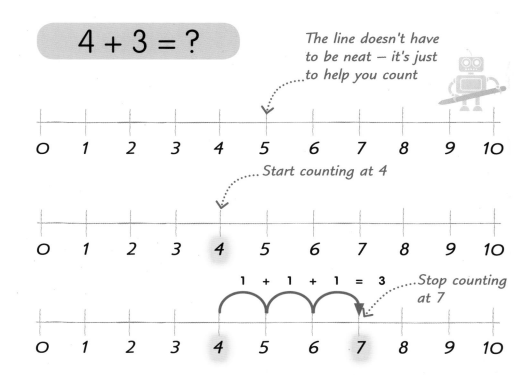

Making leaps

Some calculations involve using larger numbers. We can still use a number line, we just have to make bigger jumps to find the answer.

Jump 2 lots of 10

$$10 + 10 = 20$$

The answer is 70

| O | 10 | 20 | 30 | 40 | 50 | 60 | 70 | 80 | 90 | 100 |

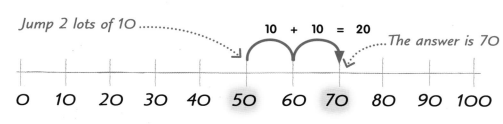

1 Let's use a number line with larger numbers to find 20 + 50.

2 Starting with the bigger number, we just have to jump 2 lots of 10 along our number line. The answer is 70.

3 So, 20 + 50 = 70

Adding with a number grid

To add numbers to 100, you can also use a number grid, or 100 square. This shows the numbers from 1 to 100 in rows of 10. You can do calculations by jumping from square to square.

Number grids are useful for calculations with numbers to 100 which are tricky to work out on a number line.

1 Look at this number grid. We can use it to add numbers in two stages. To add 10, we simply jump down to the next row, because there are 10 numbers in each row.

2 To add 1, we just jump 1 square to the right. When we get to the end of a row, we move down to the next row and carry on counting from left to right.

$$56 + 26 = ?$$

1	2	3	4	5	6	7	8	9	10
11	12	13	14	15	16	17	18	19	20
21	22	23	24	25	26	27	28	29	30
31	32	33	34	35	36	37	38	39	40
41	42	43	44	45	46	47	48	49	50
51	52	53	54	55	56	57	58	59	60
61	62	63	64	65	66	67	68	69	70
71	72	73	74	75	76	77	78	79	80
81	82	83	84	85	86	87	88	89	90
91	92	93	94	95	96	97	98	99	100

3 Let's add 56 and 26 using this number grid.

4 The addition starts with 56, so let's mark it on the grid.

5 There are 2 groups of 10 in 26, so we need to jump down 2 rows. This takes us to 76.

6 Now we add the 6 ones from our 26 by jumping 6 squares to the right. This takes us to 82.

7 So, 56 + 26 = 82

$$56 + 26 = 82$$

Addition facts

An addition fact is a simple calculation that you remember without having to work it out. Your teacher might also call this a number bond or an addition pair. Knowing simple addition facts will help you with harder calculations.

$0 + 10 = \textbf{10}$

$1 + 9 = \textbf{10}$

$2 + 8 = \textbf{10}$

$3 + 7 = \textbf{10}$

$4 + 6 = \textbf{10}$

$5 + 5 = \textbf{10}$

$6 + 4 = \textbf{10}$

$7 + 3 = \textbf{10}$

$8 + 2 = \textbf{10}$

$9 + 1 = \textbf{10}$

$10 + 0 = \textbf{10}$

Compare this fact with the last one

This is like the first fact — the numbers are just in a different order

$1 + 1 = \textbf{2}$

$2 + 2 = \textbf{4}$

$3 + 3 = \textbf{6}$

$4 + 4 = \textbf{8}$

$5 + 5 = \textbf{10}$

$6 + 6 = \textbf{12}$

$7 + 7 = \textbf{14}$

$8 + 8 = \textbf{16}$

$9 + 9 = \textbf{18}$

$10 + 10 = \textbf{20}$

These facts should be easy if you know your multiplication table for 2

1 These are called the addition facts for 10, because the answer is always 10.

2 These addition facts are all doubles. We call them the addition doubles to 10 + 10. This time, the answers are different.

TRY IT OUT

Using addition facts

Can you use the addition facts for 10 and the addition doubles to 10 + 10 to work out the answers to these calculations?

Answers on page 319

1 $60 + 40 = ?$

2 $700 + 700 = ?$

3 $20 + 80 = ?$

4 $0.1 + 0.9 = ?$

5 $70 + 30 = ?$

6 $4000 + 4000 = ?$

Partitioning for addition

Adding numbers is often easier if you split them into numbers that are easier to work with and then add them up in stages. This is called partitioning. There are a few different ways to do it.

> Partitioning means breaking numbers down then adding them together in stages.

1 Let's add 47 and 35.

$$47 + 35 = ?$$

2 To help with the tricky numbers, we can put the numbers on a grid and label the columns to show their place values.

T	O		T	O		T	O
4	7	+	3	5	=	?	?

3 We start by adding the tens together and writing the answer to the right of the equals sign: 40 + 30 = 70

T	O		T	O		T	O
4	0	+	3	0	=	7	0

4 And next, we add the ones together: 7 + 5 = 12

T	O		T	O		T	O
	7	+		5	=	1	2

5 Now it's easy to recombine our two answers to get the total: 70 + 12 = 82

Recombine the tens and ones to find the total

T	O
8	2

6 By partitioning the numbers, we've found that 47 + 35 = 82

$$47 + 35 = 82$$

Partitioning using multiples of 10

Another way to partition is to split just one number, so it's easier to add on. It often helps to split one number into a multiple of 10 and another number.

1 Let's add 80 and 54.

$$80 + 54$$

2 80 is already a multiple of 10, but we can break 54 into two parts like this: 50 + 4

$$= 80 + 50 + 4$$

3 Now we can add 50 to 80 to make 130.

$$= 130 + 4$$

4 Now we just add 4 to 130 to give the answer 134.

$$= 134$$

Expanded column addition

To add together numbers that have more than two digits, we can use column addition. There are two ways to do it. The method shown here is called expanded column addition. The other method, column addition, is shown on pages 86-87.

---

1 Let's add 385 and 157 using expanded column addition.

$$385 + 157 = ?$$

2 Start by writing the two numbers out like this, with digits that have the same place value lined up one above the other. It might help you to label the place values, but you don't have to.

H	T	O
3	8	5
+ 1	5	7

Write the numbers so that the digits with the same place value are lined up like this

3 Now we're going to add each of the digits in the top row to the digits that sit beneath them in the bottom row, starting with the ones.

H	T	O
3	8	5
+ 1	5	7

Start by adding the ones together

4 First, add 5 ones and 7 ones. The answer is 12 ones. On a new line, write 1 in the tens column and 2 in the ones column.

H	T	O
3	8	5
+ 1	5	7
	1	2

Write the answer below the answer line

When we do expanded column addition, it's important to line up the digits by their place values.

5 When we add together the 8 and 5, we're actually adding 80 and 50. The answer is 130. On a new line, write 1 in the hundreds column, 3 in the tens column and zero in the ones column.

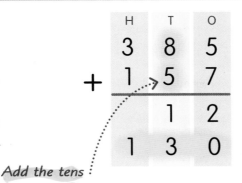

```
    H   T   O
    3   8   5
+   1   5   7
    _____
        1   2
    1   3   0
```

Add the tens together

Expanded column addition is just like partitioning – we break tricky numbers into ones, tens, and hundreds.

6 Next, we're going to add the hundreds together. We add 100 and 300 to give 400. On a new line, write 4 in the hundreds column, 0 in the tens column, and 0 in the ones column.

```
    H   T   O
    3   8   5
+   1   5   7
    _____
        1   2
    1   3   0
    4   0   0
```

Add the hundreds together

7 Now we have added the digits in the bottom row to the digits in the top row, we add the three lines in our answer together:
12 + 130 + 400 = 542

```
    H   T   O
    3   8   5
+   1   5   7
    _____
        1   2
    1   3   0
+   4   0   0
    _____
    5   4   2
```

Add the three lines in the answer together

8 So, 385 + 157 = 542

385 + 157 = 542

TRY IT OUT

Add it up

Now you have learned this useful method for adding difficult numbers, why don't you give these calculations a go?

1 547 + 276 = ?

2 948 + 642 = ?

3 7256 + 4715 = ?

Answers on page 319

Column addition

Now we're going to look at another method of column addition. This is quicker than expanded column addition (pages 84-85) because instead of writing ones, tens, and hundreds on separate lines, we put them all on one line.

Once you understand how to do column addition, you can use it for any addition calculation involving large numbers.

1 Let's use column addition to add 2795 and 4368.

$$4368 + 2795 = ?$$

2 Start by writing both numbers on a place-value grid, with the larger number above the smaller number. If you need to, label the columns.

Th	H	T	O
4	3	6	8
2	7	9	5

Place the larger number above the smaller one

3 Now we're going to add each number in the bottom row to the number that sits above it in the top row, starting with the ones.

Th	H	T	O
4	3	6	8
2	7	9	5

Start by adding the ones

4 First, add 5 to 8. The answer is 13. Write the 3 in the ones column. The 1 stands for 1 ten, so we carry it over into the tens column to add on later.

Th	H	T	O
4	3	¹6	8
2	7	9	5
			3

Carry the 1 from 13 into the tens column to add on at the next step

5 Next, we add 9 tens to 6 tens. The answer is 15 tens. Add on the 1 ten we carried over from the ones addition to make 16 tens. Write the 6 in the tens column and carry the 1 to the hundreds column.

Th	H	T	O
4	¹3	¹6	8
2	7	9	5
		6	3

Add the carried 1 ten to 15 tens, to make 16 tens

6 Now we add 7 hundreds to 3 hundreds. The answer is 10 hundreds. Add on the 1 hundred we carried over to make 11 hundreds. Write a 1 in the hundreds column and carry the other 1 to the thousands column.

Th	H	T	O
¹4	¹3	¹6	8
+ 2	7	9	5
	1	6	3

Add the carried 1 hundred to the 10 hundreds to make 11 hundreds

7 Finally, we can add the thousands. Add 2 thousands to 4 thousands. The answer is 6 thousands. Add on the 1 thousand carried over to make 7 thousands. Write the 7 in the thousands column.

Th	H	T	O
¹4	¹3	¹6	8
+ 2	7	9	5
7	1	6	3

The total of the numbers in the thousands column is less than 10, so we don't carry any numbers

8 So, 4368 + 2795 = 7163

$$4368 + 2795 = 7163$$

Adding decimals

We add decimals in the same way as we add whole numbers – we just make sure that digits of the same value are lined up underneath each other. Let's add 38.92 and 5.89.

1 First, write the larger number above the smaller number, making sure to line up the decimal points. Add another decimal point on the bottom row. If you need to, label the columns to show the place value of each.

T	O	$\frac{1}{10}$	$\frac{1}{100}$
3	8 . 9	2
+	5 . 8	9
	.	

2 Now we can find the total just as we do with whole numbers.

T	O	$\frac{1}{10}$	$\frac{1}{100}$
¹3	¹8 . 9	2
+	5 . 8	9
4	4 . 8	1

3 So, 38.92 + 5.89 = 44.81

TRY IT OUT

Can you do it?
Now you've seen how to do column addition, can you use it for these sums?

1 1639 + 6517 = ?

2 7413 + 1781 = ?

3 45.36 + 26.48 = ?

Answers on page 319

Subtraction

Subtraction is the opposite, or the inverse, of addition. There are two main ways we can think about subtraction – as taking away from a number (also called counting back) or as finding the difference between two numbers.

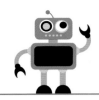

> We can use a number line for subtraction by counting either forward or back along the line.

What is subtraction?

Sometimes we reduce a number by another number. This is called subtraction as taking away. Look at these oranges. When we subtract 2 oranges from 6 oranges, there are 4 oranges left.

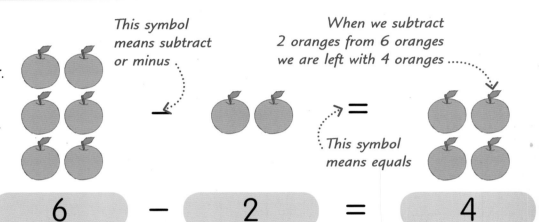

This symbol means subtract or minus

When we subtract 2 oranges from 6 oranges we are left with 4 oranges

This symbol means equals

$$6 - 2 = 4$$

Subtracting is the opposite of adding

It's easy to remember how to subtract, because it's just the opposite to addition. With addition, we add numbers on, and with subtraction we take numbers away.

To subtract we move from right to left

To add, we move from left to right

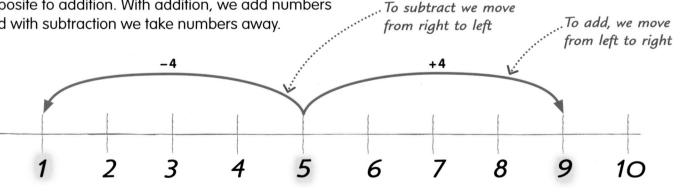

−4   +4

0   1   2   3   4   5   6   7   8   9   10

1 Subtraction
Let's use this number line to subtract 4 from 5. This takes us 4 steps back along the number line to the number 1.

2 Addition
Here, the 4 has been added to 5, and the answer is 9. We have moved the same distance from 5 as we did when subtracting, just in the other direction.

$$5 - 4 = 1$$

$$5 + 4 = 9$$

Subtracting as counting back

One way to think of subtraction is called counting back. When we subtract one number from another, we are just counting back from the first number by a number of steps that's equal to the second number.

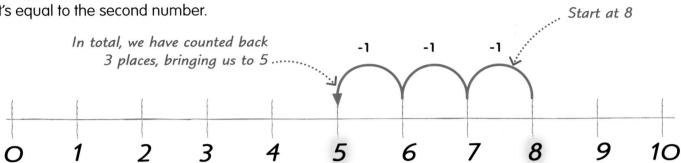

In total, we have counted back 3 places, bringing us to 5 ·········

Start at 8

1 Look at the calculation 8 – 3 on this number line.

2 To subtract 3 from 8, first we find 8, then count back 3 places. This takes us to 5.

3 So, 8 – 3 = 5

$$8 - 3 = ?$$

$$8 - 3 = 5$$

Subtracting as finding the difference

We can also think of subtraction as finding the difference between two numbers. When we are asked to find the difference, we are really just finding how many steps it takes to count from one to the other.

Then we count how many places we have to move to reach the first number.

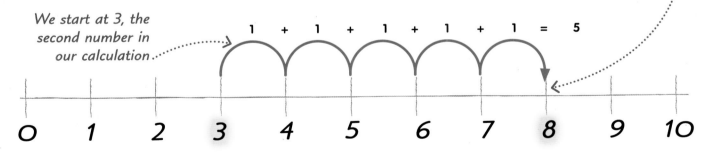

We start at 3, the second number in our calculation ·········

1 To find the difference between two numbers, we can count up a number line. Let's take another look at the calculation 8 – 3.

2 All we have to do is find 3 on the number line and see how many jumps it takes to get to 8. It takes 5 jumps.

3 So, 8 – 3 = 5

$$8 - 3 = ?$$

$$8 - 3 = 5$$

Subtraction facts

There are some simple facts that you can learn for subtraction to make tricky calculations much easier. When you've learned them, you'll be able to apply them to other calculations.

These subtraction facts are the opposite, or inverse, of the addition facts we looked at on page 82.

10 − 0 = **10**

10 − 1 = **9**

10 − 2 = **8**

10 − 3 = **7**

10 − 4 = **6**

10 − 5 = **5**

10 − 6 = **4**

10 − 7 = **3**

10 − 8 = **2**

10 − 9 = **1**

10 − 10 = **0**

Compare this fact with the last one

This is similar to the first fact — they are part of the same family of facts

2 − 1 = **1**

4 − 2 = **2**

6 − 3 = **3**

8 − 4 = **4**

10 − 5 = **5**

12 − 6 = **6**

14 − 7 = **7**

16 − 8 = **8**

18 − 9 = **9**

20 − 10 = **10**

These facts are the inverses of the doubles we looked at on page 82

1 These are the subtraction facts for 10. As the number we subtract gets larger, the difference between the two numbers gets smaller.

2 Here's another set of subtraction facts. This time, the second number in each calculation is half of the first number.

TRY IT OUT

Using subtraction facts

Can you use the subtraction facts above to work out the answers to these calculations?

Answers on page 319

1 1000 − 200 = ?

2 120 − 60 = ?

3 140 − 70 = ?

4 100 − 30 = ?

5 0.1 − 0.08 = ?

6 0.4 − 0.2 = ?

Partitioning for subtraction

Subtracting numbers is often simpler if you split them into numbers that are easier to work with and then subtract them in stages. This is called partitioning. We usually partition just the number being subtracted.

1 Let's subtract 25 from 81 by partitioning the number 25.

$$81 - 25 = ?$$

2 To help with the tricky numbers, we can put the numbers on a grid and label the columns to show their place values.

T	O		T	O		T	O
8	1	−	2	5	=	?	?

3 First, we subtract the tens from 81: 81 − 20 = 61

T	O		T	O		T	O
8	1	−	2	0	=	6	1

4 Next, we subtract the ones from the remaining 61: 61 − 5 = 56

T	O		T	O		T	O
6	1	−		5	=	5	6

5 By splitting the calculation into two easy steps, we've found that: 81 − 25 = 56

$$81 - 25 = 56$$

TRY IT OUT

Partitioning practice

There were 463 flowers in the field, and Tessa picked 86 of the flowers. How many were left in the field?

1 To work out the answer, we can do a subtraction calculation.

2 There were 463 flowers and 86 were taken away, so the calculation you need to do is: 463 − 86

3 Try partitioning the number 86 into tens and ones, and subtract it in stages from 463.

Answer on page 319

Subtracting with a number line

We have already seen that a number line can help us with simple subtraction. If we use what we know about partitioning, we can also use a number line to tackle more difficult calculations.

When you use a number line for subtraction, it doesn't matter if you count down from the first number or up from the second number, the answer will be the same.

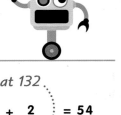

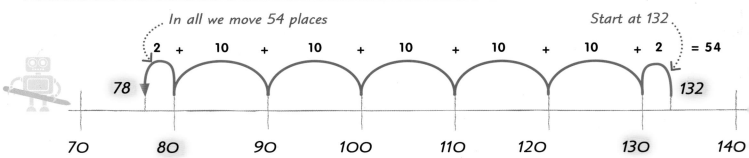

1 **Counting back**
Let's use a number line for 132 – 54. To make it easy to move along the line, we're going to partition 54 into three parts.

2 Starting from 132, we count back 2 to 130. Next, we move 50 by making 5 jumps of 10 each, taking us to 80. Finally, we move another 2 places.

3 In all, we've moved 54 places, and we've arrived at 78. So, 132 – 54 = 7

$$132 - 54 = ?$$

$$132 - 54 = 78$$

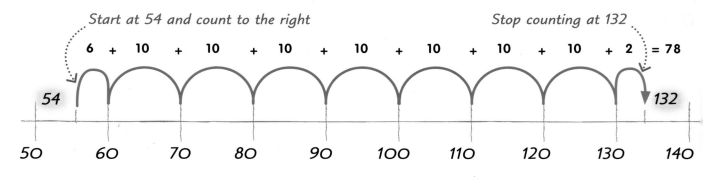

1 **Counting up**
Remember, we can also subtract by counting up. This is called finding the difference. Let's look again at 132 – 54.

2 This time, we're going to start at 54, the second number in our subtraction calculation, and count up until we get to the first number, 132.

3 First, we count up 6 places to 60. Then we take 7 jump: of 10, and finally another jump of 2. In all, we've moved 78 places. So, 132 – 54 = 78

Shopkeeper's addition

People who work in shops often need to work out quickly how much change to give a customer. They often count up in their heads to help them work out the correct change. This method of subtracting is called shopkeeper's addition.

1 Peter's groceries cost £7.35, and he pays with a £10 note. How much change is he due? We can write this as £10.00 – £7.35

$$£10.00 - £7.35 = ?$$

2 First, let's add 5p to get £7.40.

£7.35 + £0.05 = £7.40

3 Next, we add 60p to take us to £8.

£7.40 + £0.60 = £8.00

4 Now, we can add £2 to take us up to £10.

£8.00 + £2.00 = £10.00

5 Finally, we combine the amounts we've added to find the total difference: £0.05 + £0.60 + £2.00 = £2.65

£7.35 + £2.65 = £10.00

6 So, Peter is due £2.65 change from his £10 note.

$$£10.00 - £7.35 = £2.65$$

TRY IT OUT

Be the shopkeeper

Can you use the method we've learned to work out the change for these bags of shopping?

Answers on page 319

1 Cost **£3.24**
Paid for with a £10 note.

2 Cost **£17.12**
Paid for with a £20 note.

3 Cost **£59.98**
Paid for with two £50 notes.

Expanded column subtraction

To find the difference between numbers with more than two digits, we can use column subtraction. The method shown here, called expanded column subtraction, is useful if you find the ordinary column subtraction (shown on pages 96-97) difficult.

1 Let's think of the calculation 324 − 178 as finding the difference between 324 and 178.

$$324 - 178 = ?$$

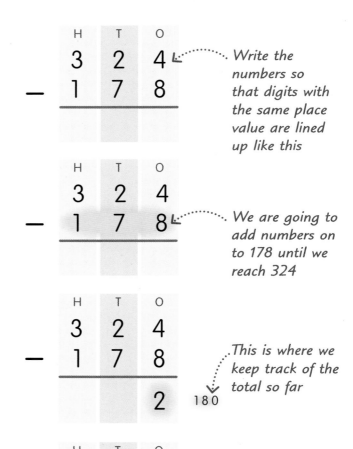

2 Start by writing the two numbers out like this, with digits that have the same place value lined up one above the other. It might help you to label the place values, but you don't have to.

> Write the numbers so that digits with the same place value are lined up like this

3 Now we're going to add numbers that are easy to work with to 178 until we get to 324.

> We are going to add numbers on to 178 until we reach 324

4 First, we add on ones that will take 178 up to the nearest multiple of ten. Adding 2 to 178 makes 180. Write 2 in the ones column. Keep track of the total, by writing 180 on the right.

> This is where we keep track of the total so far

5 Next, we add tens. Adding 20 to 180 makes 200, the nearest multiple of 100. Write the 2 in the tens column and the 0 in the ones column. Write the new total on the right.

> Adding 20 to 180 takes the total up to 200

6 Now, we add hundreds. Adding 100 takes us from 200 up to 300. Write the 1 in the hundreds column and the zeros in the tens and ones columns. Write the new total on the right.

H	T	O	
3	2	4	
− 1	7	8	
		2	180
	2	0	200
1	0	0	300

Adding 100 to 200 takes the total up to 300

7 Now we just need to add the 24 that will take the total from 300 to 324. Write the 2 in the tens column and the 4 in the ones column.

H	T	O	
3	2	4	
− 1	7	8	
		2	180
	2	0	200
1	0	0	300
+	2	4	324

Adding 24 to 300 takes the total up to 324

8 Finally, we need to find the total of all the numbers that we added on: 2 + 20 + 100 + 24 = 146

H	T	O	
3	2	4	
− 1	7	8	
		2	180
	2	0	200
1	0	0	300
+	2	4	324
1	4	6	

Find the total of the numbers we've added on

9 So, 324 − 178 = 146

$$324 - 178 = 146$$

We arrived at our answer by adding ones, tens, and hundreds in steps, like shopkeeper's addition (page 93).

TRY IT OUT

Find the difference

Can you use expanded column subtraction to find the difference between these numbers?

Answers on page 319

1 283 − 76 = ?

2 817 − 394 = ?

3 9425 − 5832 = ?

Column subtraction

Using column subtraction is an even quicker way of subtracting large numbers than expanded column subtraction (see pages 94-95). It looks tricky to subtract as we go, but we can exchange numbers with other columns to help us.

1 Let's subtract 767 from 932 using column subtraction.

$$932 - 767 = ?$$

2 Start by writing the two numbers out like this, with digits that have the same place value lined up one above the other. It might help you to label the place values, but you don't have to.

```
   H  T  O
   9  3  2
-  7  6  7
```

Write the numbers so that the digits with the same place value are lined up like this

3 Now we are going to subtract each of the digits on the bottom row from the digit above it on the top row, starting with the ones.

```
   H  T  O
   9  3  2
-  7  6  7
```

First we're going to subtract the ones

4 We can't subtract 7 ones from 2 ones here, so let's exchange 1 ten from the tens column for 10 ones. Write a little 1 next to the 2 in the ones column to show that we now have 12 ones.

```
   H  T  O
   9  3  ¹2
-  7  6  7
```

We can't subtract 7 ones from 2 ones, so we exchange 1 ten for 10 ones

5 Change the 3 in the tens column into a 2 to show that we have exchanged a ten.

```
   H   T   O
   9   ²3  ¹2
-  7   6   7
```

Change this from 3 tens to 2 tens because we exchanged 1 ten for 10 ones

6 Now we can subtract 7 ones from 12 ones instead. The answer is 5 ones. Write the 5 in the ones column.

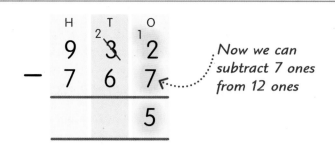

Now we can subtract 7 ones from 12 ones

7 Next, we subtract the tens. We can't subtract 6 tens from 2 tens so we need to exchange one of the hundreds for 10 tens. Write a 1 next to the 2 in the tens column to show that we now have 12 tens.

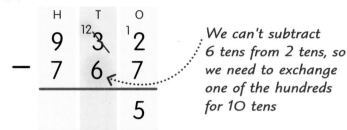

We can't subtract 6 tens from 2 tens, so we need to exchange one of the hundreds for 10 tens

8 Change the 9 in the hundreds column into an 8 to show that we have just exchanged one of the hundreds for 10 tens.

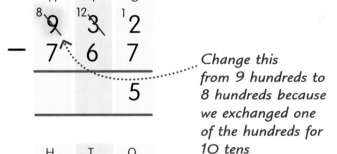

Change this from 9 hundreds to 8 hundreds because we exchanged one of the hundreds for 10 tens

9 Now we can subtract 6 tens from 12 tens. The answer is 6 tens. Write the 6 in the tens column.

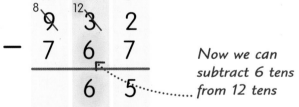

Now we can subtract 6 tens from 12 tens

10 Finally, we need to subtract 7 hundreds from 8 hundreds, leaving 1 hundred. Write the 1 in the hundreds column.

Now we can subtract 7 hundreds from 8 hundreds

11 So, 932 − 767 = 165

$$932 - 767 = 165$$

When we need to subtract a larger amount from a smaller amount, we exchange 1 ten, hundred, or thousand from the column to the left.

Multiplication

There are two main ways to think about how multiplication works. We can think of it as putting together, or adding, lots of quantities of the same size. We can also think of it as changing the scale of something – we'll look at this on page 100.

What is multiplication?

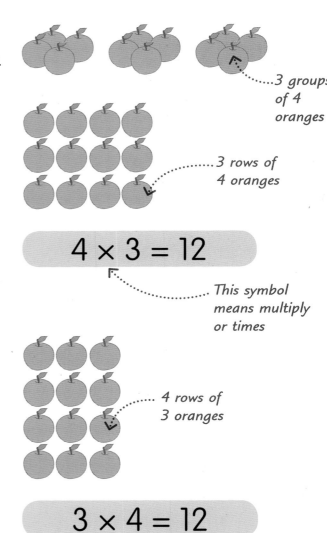

3 groups of 4 oranges

3 rows of 4 oranges

This symbol means multiply or times

4 rows of 3 oranges

The result of a multiplication is called the product

1 Look at these oranges. There are 3 groups of 4 oranges. Let's find out how many there are altogether.

2 To make them easier for us to count, let's arrange the 3 groups of 4 oranges into 3 rows of 4. We call this arrangement an array. Now it's easier for us to count them up.

3 If we count up the oranges, we can see that there are 12 altogether. We can write this as a multiplication calculation like this: 4 × 3 = 12

$$4 \times 3 = 12$$

4 Now let's line up some oranges into 4 rows of 3 instead. How many are there in total? Is it a different number of oranges to when we had 3 rows of 4 oranges?

5 If we count the oranges up, we can see that there are still 12 altogether. We can write this as a multiplication calculation too: 3 × 4 = 12

$$3 \times 4 = 12$$

6 So, 4 × 3 and 3 × 4 both give us the same total. It doesn't matter which order you multiply numbers in, the total will be the same. This means we can say that multiplication is commutative.

Multiplication as repeated addition

We can think of multiplication as adding together more than one quantity of the same size. We call this repeated addition. To multiply two numbers, we just have to add one number in the calculation to itself the number of times of the other number.

=

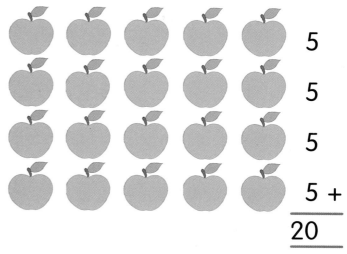

5
5
5
5 +
20

1 Let's work out the answer to the calculation 5 × 4 using some apples. We want to multiply 5 by 4, so let's look at 4 rows of 5 apples to help us find the answer.

$$5 \times 4 = ?$$

2 To work out how many apples there are in total, we just have to add 4 lots of 5: 5 + 5 + 5 + 5 = 20

3 So, using repeated addition, we can see that 5 × 4 = 20

$$5 \times 4 = 20$$

TRY IT OUT

Multiplication challenge

Here are some examples of repeated addition. Can you write them as a multiplication calculation and work out the answer?

Answers on page 319

1 6 + 6 + 6 + 6 = ?

2 8 + 8 + 8 + 8 + 8 + 8 + 8 = ?

3 9 + 9 + 9 + 9 + 9 + 9 = ?

4 13 + 13 + 13 + 13 + 13 = ?

It doesn't matter which order you multiply numbers in – the total will be the same.

Multiplication as scaling

Repeated addition is not the only way to think about multiplication. When we change the size of an object, we carry out a kind of multiplication called scaling. We also use scaling when we multiply with fractions.

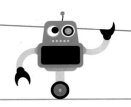

We use scaling to change the sizes of objects and to multiply with fractions.

1 Look at these three buildings. They are all different heights.

2 The second building is twice as tall as the first, so its height has been scaled up by a factor of 2. We can write this as: $10 \times 2 = 20$

3 The third building is two times taller than the second, so we can say it's been scaled up by a factor of 2. We can write this as: $20 \times 2 = 40$

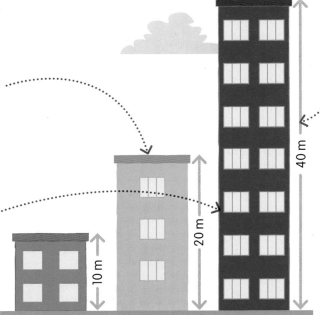

4 The third building is four times taller than the first. It has been scaled up by a factor of 4. We can write this as: $10 \times 4 = 40$

5 We could also see each building as being scaled down. The second building is half the height of the third building. We can write this using a fraction: $40 \times \frac{1}{2} = 20$

Scaling and fractions

As we've just seen, we can also scale with fractions. Multiplying with proper fractions, which are fractions less than one, makes numbers smaller, not bigger.

1 Look at this calculation. We want to multiply ¼ by ½.

2 Look at this shape. It's a quarter of a circle. To multiply a quarter by a half, we simply need to take away half of the quarter.

3 You can see that half of the quarter is one-eighth of a circle.

4 So, $\frac{1}{4} \times \frac{1}{2} = \frac{1}{8}$

$$\frac{1}{4} \times \frac{1}{2} = ?$$

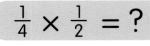
¼ of a circle

Half of ¼ of a circle

$$\frac{1}{4} \times \frac{1}{2} = \frac{1}{8}$$

Factor pairs

Two whole numbers that are multiplied together to make a third number are called factor pairs of that number. Every whole number has a factor pair, even if it's only itself multiplied by 1.

> Every whole number has at least one factor pair – the number 1 and itself.

Factor pairs for 1 to 12

Learning factor pairs is the same as learning the number facts for multiplication. Knowing these basic pairs will help you with multiplication calculations. This table shows all the factor pairs of the numbers from 1 to 12. Each pair has also been drawn as an array, like the arrays we saw on pages 98-99.

pages 98-99

Number	Factor pairs	Array
1	1 , 1	
2	1 , 2	
3	1 , 3	
4	1 , 4	
	2 , 2	
5	1 , 5	
6	1 , 6	
	2 , 3	
7	1 , 7	
8	1 , 8	
	2 , 4	
9	1 , 9	
	3 , 3	
10	1 , 10	
	2 , 5	
11	1 , 11	
12	1 , 12	
	2 , 6	
	3 , 4	

TRY IT OUT

Finding pairs

Can you find all the factor pairs for each of these numbers? Draw them out as arrays if you find it helpful.

1 14

2 60

3 18

4 35

5 24

Answers on page 319

Counting in multiples

When a whole number is multiplied by another whole number, the result is called a multiple – we looked at multiples on pages 30-31. When we're doing multiplication calculations, it helps to know how to count in multiples.

1 Counting in 2s
Look at this number line. It shows the numbers we get when we count up in twos from zero. Each number in the sequence is a multiple of 2. For example, the fourth jump takes us to 8, so $2 \times 4 = 8$

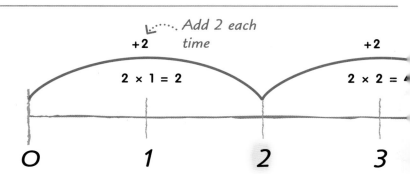

2 Counting in 3s
This number line shows the numbers we get when we start to count in multiples of three from zero. The fifth jump takes us to 15, so $3 \times 5 = 15$

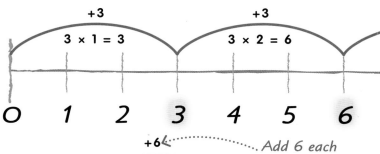

3 Counting in 6s
Now look at this number line. It shows us the first few multiples of six. The third jump takes us to 18, so we can say that $6 \times 3 = 18$

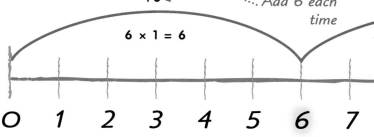

4 Counting in 8s
This number line shows us the first three multiples of 8 when we count up from zero. The second jump takes us to 16, so $8 \times 2 = 16$

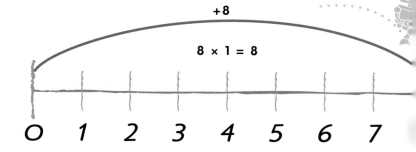

5 These number lines show us the first few multiples of 2, 3, 6, and 8. Learning to count in multiples will help us with other multiplication tables, which we'll look at on pages 104-105.

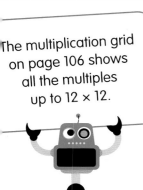

The multiplication grid on page 106 shows all the multiples up to 12 × 12.

TRY IT OUT

Find the multiples

Now you've seen the first few multiples of the numbers 2, 3, 6, and 8, can you use a number line, or count in your head, to find the next three multiples for 7, 9, and 11?

Answers on page 319

1 7, 14, 21 ...

2 9, 18, 27 ...

3 11, 22, 33 ...

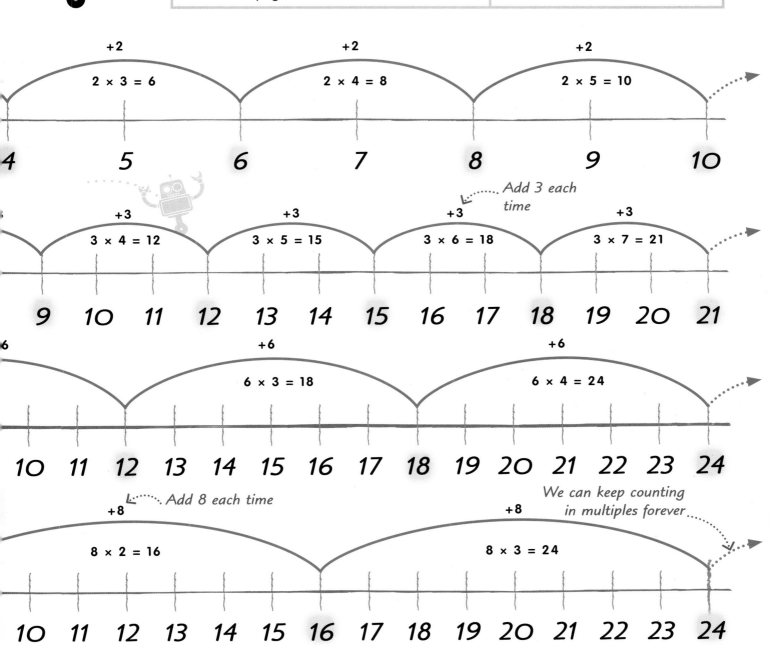

+2
2 × 3 = 6

+2
2 × 4 = 8

+2
2 × 5 = 10

4   5   6   7   8   9   10

Add 3 each time

+3
3 × 4 = 12

+3
3 × 5 = 15

+3
3 × 6 = 18

+3
3 × 7 = 21

9   10   11   12   13   14   15   16   17   18   19   20   21

+6

+6
6 × 3 = 18

+6
6 × 4 = 24

10   11   12   13   14   15   16   17   18   19   20   21   22   23   24

Add 8 each time

We can keep counting in multiples forever

+8
8 × 2 = 16

+8
8 × 3 = 24

10   11   12   13   14   15   16   17   18   19   20   21   22   23   24

Multiplication tables

The multiplication tables are really just a list of the multiplication facts about a particular number. You need to learn them – but once you know them, you'll find them very useful when you're doing other calculations.

1× table

1	×	0	=	**0**
1	×	1	=	**1**
1	×	2	=	**2**
1	×	3	=	**3**
1	×	4	=	**4**
1	×	5	=	**5**
1	×	6	=	**6**
1	×	7	=	**7**
1	×	8	=	**8**
1	×	9	=	**9**
1	×	10	=	**10**
1	×	11	=	**11**
1	×	12	=	**12**

2× table

2	×	0	=	**0**
2	×	1	=	**2**
2	×	2	=	**4**
2	×	3	=	**6**
2	×	4	=	**8**
2	×	5	=	**10**
2	×	6	=	**12**
2	×	7	=	**14**
2	×	8	=	**16**
2	×	9	=	**18**
2	×	10	=	**20**
2	×	11	=	**22**
2	×	12	=	**24**

3× table

3	×	0	=	**0**
3	×	1	=	**3**
3	×	2	=	**6**
3	×	3	=	**9**
3	×	4	=	**12**
3	×	5	=	**15**
3	×	6	=	**18**
3	×	7	=	**21**
3	×	8	=	**24**
3	×	9	=	**27**
3	×	10	=	**30**
3	×	11	=	**33**
3	×	12	=	**36**

4× table

4	×	0	=	**0**
4	×	1	=	**4**
4	×	2	=	**8**
4	×	3	=	**12**
4	×	4	=	**16**
4	×	5	=	**20**
4	×	6	=	**24**
4	×	7	=	**28**
4	×	8	=	**32**
4	×	9	=	**36**
4	×	10	=	**40**
4	×	11	=	**44**
4	×	12	=	**48**

5× table

5	×	0	=	**0**
5	×	1	=	**5**
5	×	2	=	**10**
5	×	3	=	**15**
5	×	4	=	**20**
5	×	5	=	**25**
5	×	6	=	**30**
5	×	7	=	**35**
5	×	8	=	**40**
5	×	9	=	**45**
5	×	10	=	**50**
5	×	11	=	**55**
5	×	12	=	**60**

6× table

6	×	0	=	**0**
6	×	1	=	**6**
6	×	2	=	**12**
6	×	3	=	**18**
6	×	4	=	**24**
6	×	5	=	**30**
6	×	6	=	**36**
6	×	7	=	**42**
6	×	8	=	**48**
6	×	9	=	**54**
6	×	10	=	**60**
6	×	11	=	**66**
6	×	12	=	**72**

TRY IT OUT

The 13× table

You should know your multiplication tables up to 12. Here are the first four lines of the 13x table. Can you work out the rest?

Answers on page 319

13	×	1	=	**13**
13	×	2	=	**26**
13	×	3	=	**39**
13	×	4	=	**?**
.....				

7× table

7	×	0	=	**0**
7	×	1	=	**7**
7	×	2	=	**14**
7	×	3	=	**21**
7	×	4	=	**28**
7	×	5	=	**35**
7	×	6	=	**42**
7	×	7	=	**49**
7	×	8	=	**56**
7	×	9	=	**63**
7	×	10	=	**70**
7	×	11	=	**77**
7	×	12	=	**84**

8× table

8	×	0	=	**0**
8	×	1	=	**8**
8	×	2	=	**16**
8	×	3	=	**24**
8	×	4	=	**32**
8	×	5	=	**40**
8	×	6	=	**48**
8	×	7	=	**56**
8	×	8	=	**64**
8	×	9	=	**72**
8	×	10	=	**80**
8	×	11	=	**88**
8	×	12	=	**96**

9× table

9	×	0	=	**0**
9	×	1	=	**9**
9	×	2	=	**18**
9	×	3	=	**27**
9	×	4	=	**36**
9	×	5	=	**45**
9	×	6	=	**54**
9	×	7	=	**63**
9	×	8	=	**72**
9	×	9	=	**81**
9	×	10	=	**90**
9	×	11	=	**99**
9	×	12	=	**108**

10× table

10	×	0	=	**0**
10	×	1	=	**10**
10	×	2	=	**20**
10	×	3	=	**30**
10	×	4	=	**40**
10	×	5	=	**50**
10	×	6	=	**60**
10	×	7	=	**70**
10	×	8	=	**80**
10	×	9	=	**90**
10	×	10	=	**100**
10	×	11	=	**110**
10	×	12	=	**120**

11× table

11	×	0	=	**0**
11	×	1	=	**11**
11	×	2	=	**22**
11	×	3	=	**33**
11	×	4	=	**44**
11	×	5	=	**55**
11	×	6	=	**66**
11	×	7	=	**77**
11	×	8	=	**88**
11	×	9	=	**99**
11	×	10	=	**110**
11	×	11	=	**121**
11	×	12	=	**132**

12× table

12	×	0	=	**0**
12	×	1	=	**12**
12	×	2	=	**24**
12	×	3	=	**36**
12	×	4	=	**48**
12	×	5	=	**60**
12	×	6	=	**72**
12	×	7	=	**84**
12	×	8	=	**96**
12	×	9	=	**108**
12	×	10	=	**120**
12	×	11	=	**132**
12	×	12	=	**144**

The multiplication grid

We can arrange all the numbers in the multiplication tables in a grid called a multiplication grid. The factors appear along the top of the grid and down one side. The answers are in the middle.

1 Let's use the grid to find 3 × 7.

$$3 \times 7 = ?$$

2 All we need to do is find the first factor along the top of the grid. This is 3.

3 The second factor is 7, so next we look for 7 down the side of the grid.

×	1	2	3	4	5	6	7	8	9	10	11	12
1	1	2	3	4	5	6	7	8	9	10	11	12
2	2	4	6	8	10	12	14	16	18	20	22	24
3	3	6	9	12	15	18	21	24	27	30	33	36
4	4	8	12	16	20	24	28	32	36	40	44	48
5	5	10	15	20	25	30	35	40	45	50	55	60
6	6	12	18	24	30	36	42	48	54	60	66	72
7	7	14	21	28	35	42	49	56	63	70	77	84
8	8	16	24	32	40	48	56	64	72	80	88	96
9	9	18	27	36	45	54	63	72	81	90	99	108
10	10	20	30	40	50	60	70	80	90	100	110	120
11	11	22	33	44	55	66	77	88	99	110	121	132
12	12	24	36	48	60	72	84	96	108	120	132	144

Remember, multiplication can be done in any order so you can look for a factor either along the top or down the side.

4 Finally, move along and down from the two factors until the row and column meet.

5 Our two factors, 3 and 7, meet at the box in the grid for 21.

6 So, 3 × 7 = 21

$$3 \times 7 = 21$$

Multiplication patterns and strategies

There are lots of patterns and simple strategies that will help you to learn your multiplication tables and even go beyond them. Some of the easiest to remember are shown in the table on this page.

To multiply	How to do it	Examples
×2	Double the number – that is, add it to itself.	$2 \times 11 = 11 + 11 = 22$
×4	Double the number, then double again.	$8 \times 4 = 32$, because double 8 is 16 and double 16 is 32.
×5	The ones digit of multiples of 5 follow the pattern 5, 0, 5, 0 …	The first four answers in the 5× table are **5, 10, 15,** and **20.**
	Multiply by 10 then halve the result.	$16 \times 5 = 80$, because $16 \times 10 = 160$, then halve 160 to make 80.
×9	Multiply the number by 10, then subtract the number.	$9 \times \mathbf{7} = (10 \times \mathbf{7}) - \mathbf{7} = 63$
	For calculations up to 9 × 10, you can use a method that involves counting your fingers.	To work out 3 × 9, hold your hands up with your palms facing you. Then hold down your third finger from the left. There are 2 fingers to its left and 7 to its right, so the answer is 27.
×11	To multiply the numbers 1 to 9 by 11, write the digit twice, once in the tens place and once in the ones place.	$\mathbf{4} \times 11 = \mathbf{44}$
×12	Multiply the original number by 10, then multiply it by 2, then add the two answers.	$12 \times \mathbf{3} = (10 \times \mathbf{3}) + (2 \times \mathbf{3}) = 30 + 6 = 36$

Multiplying by 10, 100, and 1000

Multiplying by 10, 100, and 1000 is straightforward. To multiply a number by 10, for example, all you have to do is shift each of its digits one place to the left on a place-value grid.

> To multiply a number by 10, we just move each of its digits one place to the left.

1 Multiplying by 10
Let's multiply 3.2 by 10. To work out the answer, we just move each digit one place to the left on the place-value grid. So, 3.2 becomes 32, ten times bigger than 3.2.

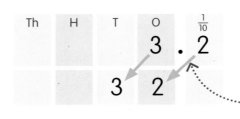

Move each digit one place to the left

2 Multiplying by 100
Let's try multiplying 3.2 by 100 this time. To multiply a number by 100, we shift each digit two places to the left. So, 3.2 becomes 320, 100 times bigger than 3.2.

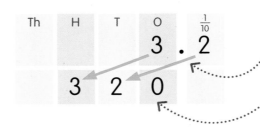

Move each digit two places to the left

Add a O as a place holder in the ones column

3 Multiplying by 1000
Now let's multiply 3.2 by 1000. To do this, we move each digit three places to the left. So, 3.2 becomes 3200, 1000 times bigger than 3.2.

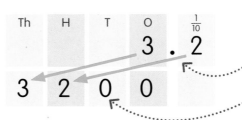

Move each digit three places to the left

Add two Os as place holders in the tens and ones columns

4 We can keep going like this for 10 000, 100 000, and even 1 000 000.

TRY IT OUT

Step to the left

Can you use the method we have shown you to work out the answers to these calculations?

Answers on page 319

1 $6.79 \times 100 = ?$

2 $48 \times 10\,000 = ?$

3 $0.072 \times 1000 = ?$

Multiplying by multiples of 10

To make multiplication calculations involving multiples of 10 easier, you can combine what you know about the multiplication tables with what you know about multiplying by 10.

To multiply a number by a multiple of 10, break the multiple into 10 and its other factor and do the calculation in steps.

1 Look at this calculation. We want to multiply 126 by 20. It looks tricky, but it's simple if you know your multiples of 10.

$$126 \times 20 = ?$$

2 Let's write 20 as 2 × 10, because multiplying by 2 and 10 are easier than multiplying by 20.

$$126 \times 2 \times 10$$

3 Now we can multiply 126 by 2. We know that 26 × 2 = 52, so we can work out that 126 × 2 = 252

$$126 \times 2 = 252$$

4 Finally, we just have to multiply 252 by 10. The answer is 2520.

$$252 \times 10 = 2520$$

5 So, 126 × 20 = 2520

$$126 \times 20 = 2520$$

TRY IT OUT

Trickier tens

Look at these calculations. Can you break down the multiples of 10 to make each calculation simpler and work out the answer?

Answers on page 319

1 $25 \times 50 = ?$

2 $0.5 \times 60 = ?$

3 $231 \times 30 = ?$

4 $43 \times 70 = ?$

5 $0.03 \times 90 = ?$

6 $824 \times 20 = ?$

Partitioning for multiplication

Just like we do for addition, subtraction, and division, we can partition numbers in a multiplication calculation in order to make it easier to find the answer.

Partitioning on a number line

We can use a number line to break up one of the numbers in a calculation into two smaller numbers that are easier to work with.

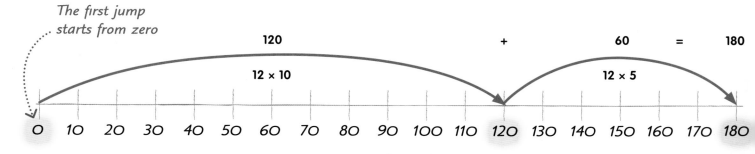

1 Let's use partitioning on a number line to answer this question: a lorry is 12 m long, and a train is 15 times longer. How long is the train?

2 To find the answer, we need to multiply the length of the lorry, which is 12 m, by 15.

3 We can partition either number in the calculation. Let's partition the number 15 into 10 and 5.

$$12 \times 15 = ?$$

The first jump starts from zero

120   +   60   =   180

12 × 10   12 × 5

O 10 20 30 40 50 60 70 80 90 100 110 120 130 140 150 160 170 180

4 First, multiply 12 by 10. The answer is 120. So, we jump up the number line from 0 to 120.

5 Next, we multiply 12 by the remaining 5. The answer is 60. So, we jump up the number line 60 from 120 to 180.

6 So, the train is 180 m long.

$$12 \times 15 = 180$$

Partitioning on a grid

We can also use a grid to help us to partition for multiplication. A grid like this is called an open array.

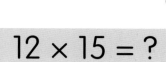

It doesn't matter which number in a calculation you choose to partition – just pick whichever is simpler to work with.

1 Let's take another look at 12×15, this time using a grid. As before, we can partition 15 into 10 and 5.

$$12 \times 15 = ?$$

2 First, draw a rectangle, like this one, where each side represents a number in the calculation. We can draw the grid roughly, without using a ruler or measuring the sides.

	10	5
12	$12 \times 10 = 120$	$12 \times 5 = 60$

3 We are partitioning 15 into 10 and 5, so we draw a line through the rectangle to show that it has been partitioned. Label the sides with 12 on one side, and 5 and 10 on the other.

4 Now we multiply the sides of each section of the grid. First, multiply 12 by 10 to get 120. Write $12 \times 10 = 120$ in the grid.

5 Next, multiply 12 by 5 to get 60. Write $12 \times 5 = 60$ in the grid.

6 Finally, we just add the two answers together: $120 + 60 = 180$

7 So, $12 \times 15 = 180$

$$12 \times 15 = 180$$

8 We can also partition this calculation without drawing a grid. We can write it like this:
$12 \times 15 = (12 \times 10) + (12 \times 5) = 120 + 60 = 180$

TRY IT OUT

Partitioning practice

Try using the number line and grid methods to work out the answers to these multiplication calculations. Which method do you prefer?

1 $35 \times 22 = ?$ **3** $26 \times 12 = ?$

2 $17 \times 14 = ?$ **4** $16 \times 120 = ?$

Answers on page 319

The grid method

We can also use a slightly different version of the open array we saw on page 111. We call it the grid method. As you get better, the grid can become simpler and you can find the answers to tricky multiplication calculations faster.

Knowing your multiplication tables and multiples of 10 will help you to get quicker at using the grid method.

1 Let's use the grid method to work out 37 × 18.

$$37 \times 18 = ?$$

2 First, draw a rectangle and label the sides with the numbers in the calculation: 37 and 18. We can draw the grid roughly, without using a ruler or measuring the sides.

37

18

Label the sides of a rectangle with the numbers in the calculation

3 Next, we partition 37 and 18 into smaller numbers that are easier to calculate with. Let's split 18 into 10 and 8, and draw a line across the rectangle between the two numbers.

Split 18 into 10 and 8

37

18 — 10 / 8

4 Now we partition 37 into 10, 10, 10, and 7. Draw lines down the rectangle between each number. Our rectangle now looks like a grid.

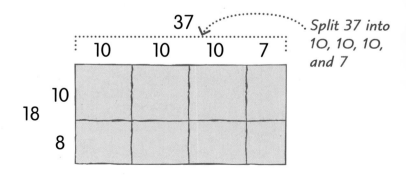

Split 37 into 10, 10, 10, and 7

5 Next, multiply the number at the top of each column by the number at the start of each row, and write the product in each box in the grid.

Multiply the number at the top of the column by the number at the start of the row

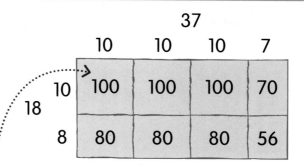

6 Finally, we simply add up all the numbers in the grid row by row and write the total at the end of each row. We get 370 and 296. Then we can add these numbers together using column addition to find the total: $370 + 296 = 666$

Find the total of each row

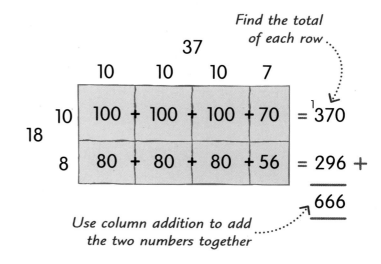

Use column addition to add the two numbers together

7 So, $37 \times 18 = 666$

$$37 \times 18 = 666$$

Quicker grid methods

When we get more confident with multiplication calculations, we can use quicker forms of the grid method. They are like the one we have just used, but have fewer steps and a simpler grid. Here are two shorter grid methods to work out 37×18.

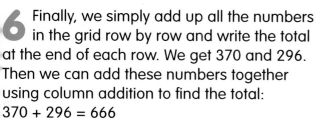

Partition the numbers into fewer chunks

Draw a simpler grid

×	30	7	
10	300	70	$= 370$
8	240	56	$= 296 +$
			666

1 If we partition the numbers in a calculation into fewer, larger chunks, we don't have to do so many calculations.

2 Once we understand what we are doing, we can draw a quick and simple grid instead of a box.

Expanded short multiplication

When one of the numbers in a multiplication calculation has more than one digit, it can help to write the numbers out in columns. There's more than one way to do this. The method shown here, called expanded short multiplication, is useful when you're multiplying a number with more than one digit by a single-digit number.

1 Let's multiply 423 by 8 using expanded short multiplication.

$$423 \times 8 = ?$$

2 Start by writing the two numbers out like this, with digits that have the same place values lined up one above the other. It might help you to label the place values, but you don't have to.

Th	H	T	O
	4	2	3
			8

×

Write the numbers so that the digits with the same place values are lined up like this

3 Now we're going to multiply each of the digits on the top row by the number 8 on the bottom row, starting with the ones.

Th	H	T	O
	4	2	3
			8

×

We're going to multiply each digit on the top row by 8

4 First, multiply 3 ones by 8 ones. The answer is 24 ones. Write 24 on the first answer row.

Th	H	T	O
	4	2	3
			8
		2	4

×

Write the answer on a row below

5 Next, we multiply the 2 tens by 8 ones. The answer is 16 tens. This is the same as 160, so we write 160 on the row beneath 24.

Th	H	T	O
	4	2	3
×			8
		2	4
	1	6	0

Multiply the tens digit by 8

6 Now we multiply the 4 hundreds by 8 ones. The answer is 32 hundreds. This is the same as 3200, so we write 3200 on the line below 160.

Th	H	T	O
	4	2	3
×			8
		2	4
	1	6	0
3	2	0	0

Multiply the hundreds digit by 8

7 Finally, we just need to add together our three answers to get the final answer: 24 + 160 + 3200 = 3384

Th	H	T	O
	4	2	3
×			8
		2	4
	1	6	0
+ 3	2	0	0
3	3	8	4

Add the three lines in the answer together

8 So, 423 × 8 = 3384

423 × 8 = 3384

TRY IT OUT

Stretch yourself

If a single spider has 8 legs, how many legs do 384 spiders have?

1 We can use expanded short multiplication to work out the answer. We simply need to multiply 8 by 384.

2 All we need to do is multiply each digit of 384 by 8 then add the answers together.

As you multiply numbers with more digits, you'll need to add extra rows to your answer.

Answer on page 319

Short multiplication

Now we're going to look at another method of short multiplication. This is quicker than expanded short multiplication (which we looked at on pages 114-15) because instead of writing the ones, tens, and hundreds in our answer on separate lines and then adding them up, we put them all on one line.

1 Let's use short multiplication to multiply 736 by 4.

$$736 \times 4 = ?$$

2 Start by writing the two numbers out like this, with digits that have the same place values lined up one above the other. It might help you to label the place values, but you don't have to.

Th	H	T	O
	7	3	6
×			4

Write the numbers so that the digits with the same place values are lined up like this

3 Now we're going to multiply each of the digits on the top row by the number 4 on the bottom row.

Th	H	T	O
	7	3	6
×			4

We're going to multiply each digit on the top row by 4

4 First, multiply 6 ones by 4 ones. The answer is 24 ones. Write the 4 in the ones column. The 2 stands for 2 tens, so we carry it over into the tens column to add on at the next stage.

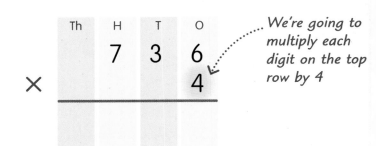
Multiply 6 ones by 4 ones

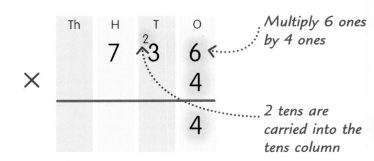
2 tens are carried into the tens column

5 Next, we multiply 3 tens by 4 ones. The answer is 12 tens. Add on the 2 tens we carried over from the ones multiplication to make 14 tens. Write the 4 in the tens column, and carry the 1 to the hundreds column.

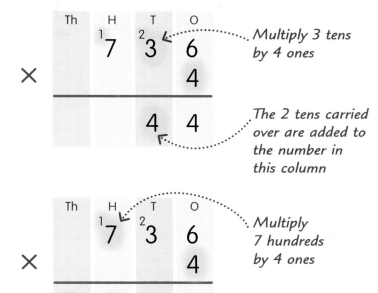

Multiply 3 tens by 4 ones

The 2 tens carried over are added to the number in this column

6 Now we multiply 7 hundreds by 4 ones. The answer is 28 hundreds. Add on the 1 hundred we carried over from the tens multiplication to make 29 hundreds. Write the 9 in the hundreds column and the 2 in the thousands column.

Multiply 7 hundreds by 4 ones

The 1 hundred carried over is added to the number in this column

7 So, 736 × 4 = 2944

$$736 \times 4 = 2944$$

TRY IT OUT

Test your skills

Can you use short multiplication to work out the answers to these calculations? For the numbers that have four digits, just add an extra column to your answer for the thousands.

1 295 × 8 = ?

2 817 × 5 = ?

3 2739 × 3 = ?

4 4176 × 4 = ?

5 6943 × 9 = ?

Once you understand how to do short multiplication, you can use it for multiplying any number with more than one digit by a number with just one digit.

Answers on page 319

Expanded long multiplication

When we need to multiply two numbers that both have two or more digits, we can use a method called long multiplication. There are two main ways to do it. The method shown here is called expanded long multiplication. The other method, called long multiplication, is shown on pages 120-23.

1 Let's multiply 37 by 16 using expanded long multiplication.

2 Start by writing the two numbers out like this, with digits that have the same place values lined up one above the other. It might help you to label the place values, but you don't have to.

3 Now we're going to multiply each of the digits on the top row by each of the digits on the bottom row. We'll start by multiplying all of the digits on the top row by 6 ones.

4 First, multiply 7 ones by 6 ones. The answer is 42 ones. On a new line, write 4 in the tens column and 2 in the ones column.

5 Next, multiply 3 tens by 6 ones. The answer is 18 tens, or 180. On a new line, write 1 in the hundreds column, 8 in the tens column, and 0 in the ones column.

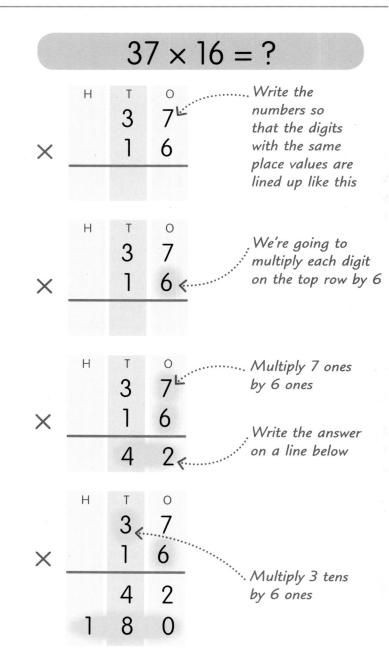

$$37 \times 16 = ?$$

Write the numbers so that the digits with the same place values are lined up like this

We're going to multiply each digit on the top row by 6

Multiply 7 ones by 6 ones

Write the answer on a line below

Multiply 3 tens by 6 ones

6 Now we're going to multiply all the digits on the top row by 1 ten and continue to write the answers below.

H	T	O
	3	7
×	1	6
	4	2
1	8	0

We're going to multiply each digit in the top row by 1 ten

7 First, multiply 7 ones by 1 ten. The answer is 7 tens, or 70. On another new line, write 7 in the tens column and 0 in the ones column.

H	T	O
	3	7
×	1	6
	4	2
1	8	0
	7	0

Multiply 7 ones by 1 ten

8 Next, multiply 3 tens by 1 ten. The answer is 30 tens, or 300, because we are multiplying 30 by 10. On a new line, write 3 in the hundreds column, 0 in the tens column, and 0 in the ones column.

H	T	O
	3	7
×	1	6
	4	2
1	8	0
	7	0
3	0	0

Multiply 3 tens by 1 ten

9 Now we have multiplied all the digits on the top line by all the digits on the second line, we add all four lines in our answer together:
42 + 180 + 70 + 300 = 592

10 So, 37 × 16 = 592

Add the four answers together

$$37 × 16 = 592$$

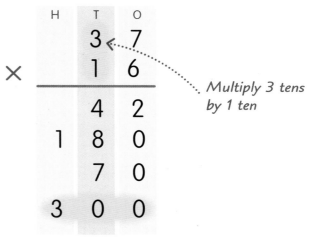

H	T	O
	3	7
×	1	6
1		
	4	2
1	8	0
	7	0
+ 3	0	0
5	9	2

When we add 4 tens, 8 tens, 7 tens, and O tens, we get 19 tens, so carry the 1 into the hundreds column

Long multiplication

Now we're going to look at another method of long multiplication (which we also looked at on pages 118-19). It's another way to multiply numbers that have two or more digits, but this method is quicker.

> Once you understand how to do long multiplication, you can use it for multiplying two numbers with any number of digits.

1 Let's multiply 86 by 43 using long multiplication.

$$86 \times 43 = ?$$

2 Start by writing the two numbers out like this, with digits that have the same place values lined up one above the other. It might help you to label the place values, but you don't have to.

Write the numbers so that the digits with the same place value are lined up like this

3 Now we're going to multiply each of the digits on the top row by each of the digits on the bottom row. Start by multiplying all the numbers on the top row by 3 ones.

We're going to multiply each digit on the top row by 3 ones

4 First, multiply 6 ones by 3 ones. The answer is 18 ones. On a new line, write 8 in the ones column. The 1 stands for 1 ten, so we carry it over into the tens column to add on at the next stage.

Multiply 6 ones by 3 ones

1 ten carried into the tens column

5 Next, multiply 8 tens by 3 ones. The answer is 24 tens. Add the 1 ten that we carried over from the ones multiplication to make 25 tens, or 250. Write the 2 in the hundreds column and the 5 in the tens column.

Multiply 8 tens by 3 ones

The 1 ten carried over is added to the number put in this column

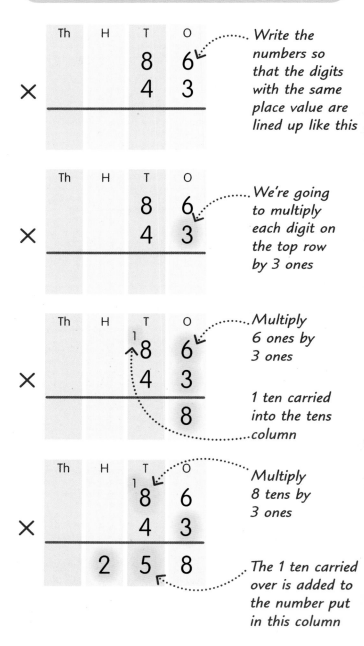

6 Now we're going to multiply all the digits on the top row by 4 tens and write the answers on a new line.

Th	H	T	O
		¹8	6
×		4	3
	2	5	8

Multiply 8 tens and 6 ones by 4 tens

7 When we multiply by this 4, we're actually multiplying by 40, which is 10 times 4. So, first we put a 0 in the ones column on a new line as a place holder.

Th	H	T	O
		¹8	6
×		4	3
	2	5	8
			0

This 4 means 4 tens or 40

Put a 0 on a new line in the ones column

8 Now multiply 6 ones by 4 tens. The answer is 24 tens. Write the 4 in the tens column and carry the 2 into the hundreds column to add on at the next stage.

Th	H	T	O
	²	¹8	6
×		4	3
	2	5	8
		4	0

Multiply 6 ones by 4 tens

2 carried into the hundreds column

9 Next, multiply 8 tens by 4 tens. The answer is 32 hundreds. Add the 2 hundreds that we carried over to make 34 hundreds. Write the 4 in the hundreds column and the 3 in the thousands column.

Th	H	T	O
	²	¹8	6
×		4	3
	2	5	8
3	4	4	0

Multiply 8 tens by 4 tens

The 2 hundreds carried over is added to the number put in this column

10 Now we have multiplied all the digits on the top row by all the digits on the bottom row, we add the two lines on our answer together: 258 + 3440 = 3698

Add the two lines in the answer together

Th	H	T	O
	²	¹8	6
×		4	3
	2	5	8
+ 3	4	4	0
3	6	9	8

The final stage of our calculation involves column addition. We looked at this on pages 86-87.

11 So, 86 × 43 = 3698

86 × 43 = 3698

More long multiplication

When we need to multiply a number that has more than two digits by a two-digit number, we can also use long multiplication. It may look trickier with such a large number, but all we need to do is use more steps.

1 Let's multiply 7242 by 23.

2 Start by writing the two numbers out like this, with digits that have the same place values lined up one above the other. Now we're going to multiply each of the digits on the top row by each of the digits on the bottom row, starting with the ones.

HTh	TTh	Th	H	T	O
		7	2	4	2
×				2	3

We're going to multiply each digit on the top row by 3 ones

3 First, multiply the 2 ones by 3 ones. The answer is 6 ones. On a new line, write 6 in the ones column.

HTh	TTh	Th	H	T	O
		7	2	4	2
×				2	3
					6

Multiply 2 ones by 3 ones

4 Next, multiply 4 tens by 3 ones. The answer is 12 tens, or 120. Write 2 in the tens column. The 1 stands for 1 hundred, so we carry it over into the hundreds column to add on at the next stage.

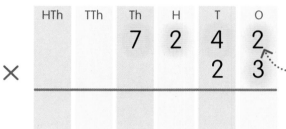

1 carried into the hundreds column

Multiply 4 tens by 3 ones

5 Now multiply 2 hundreds by 3 ones. The answer is 6 hundreds. Add the 1 hundred that we carried over from the tens multiplication to make 7 hundreds. Write the 7 in the hundreds column.

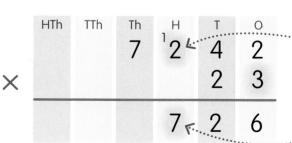

Multiply 2 hundreds by 3 ones

The 1 carried over is added to the number put in this column

6 Next, multiply 7 thousands by 3 ones. The answer is 21 thousands. Write 1 in the thousands column and 2 in the ten thousands column.

HTh	TTh	Th	H	T	O
		7	¹2	4	2
				2	3
	2	1	7	2	6

Multiply 7 thousands by 3 ones

7 Now we're going to multiply all the digits on the top row by 2 tens and write the answers on a new line. When we multiply by the 2 tens, we're actually multiplying by 20, which is 10 times 2. So, first we put a 0 in the ones column on the new line as a place holder.

HTh	TTh	Th	H	T	O
		7	¹2	4	2
				2	3
	2	1	7	2	6
					0

This 2 means 2 tens, or 20

Put a 0 on a new line in the ones column

8 Next, we multiply each of the digits in the top row by the 2 tens, in the same way that we did when we multiplied the top row by 3. The answer on the bottom line is 144 840.

HTh	TTh	Th	H	T	O
		7	¹2	4	2
				2	3
	2	1	7	2	6
1	4	4	8	4	0

Multiply each digit on the top row by 2 tens

9 Now we have multiplied all the digits on the top row by all the digits on the bottom row, we use column addition to add the two lines in our answer together: 21 726 + 144 840 = 166 566

Add the two answers together

HTh	TTh	Th	H	T	O
		7	¹2	4	2
				2	3
	2	¹1	7	2	6
1	4	4	8	4	0
1	6	6	5	6	6

10 So, 7242 × 23 = 166 566

$$7242 \times 23 = 166\,566$$

Multiplying decimals

We can use long multiplication to multiply decimals. It might look tricky, but really it's just as simple as multiplying any other number. All we have to do is make sure we carefully line up the decimal point in the answer line with the decimal point in the question.

When multiplying with decimals, it helps to estimate the answer first, so you can see at the end if you've made a mistake.

1 Let's multiply 6.3 by 52.

$$6.3 \times 52 = ?$$

2 First, write the number with the decimal number above the whole number. We don't need to line up the numbers according to their place values. Write a decimal point on a new line, below the decimal point in the question.

$$\begin{array}{r} 6\,.\,3 \\ \times\quad 5\quad 2 \\ \hline . \end{array}$$

We don't need to line up the numbers by place value

Line up this decimal point with the one in the question

3 Now we're going to multiply each of the digits on the top row by each digit on the bottom row. Start by multiplying all the digits by 2.

$$\begin{array}{r} 6\,.\,3 \\ \times\quad 5\quad 2 \\ \hline . \end{array}$$

We'll multiply each digit on the top row by 2

4 First, multiply 3 by 2. The answer is 6. Write 6 in the first column.

$$\begin{array}{r} 6\,.\,3 \\ \times\quad 5\quad 2 \\ \hline .\ 6 \end{array}$$

Multiply 3 by 2

Write the 6 here

5 Next, multiply 6 by 2. The answer is 12. Write 2 in the next column to the left of the decimal point, and 1 in the next column.

$$\begin{array}{r} 6\,.\,3 \\ \times\quad 5\quad 2 \\ \hline 1\ 2\,.\,6 \end{array}$$

Multiply 6 by 2

6 Now we're going to multiply all the digits on the top row by 5 and write the answers on a new line. Write a decimal point on this new line, in line with the other decimal points.

$$\begin{array}{r} 6 \cdot 3 \\ \times \quad 5 \quad 2 \\ \hline 1 \quad 2 \cdot 6 \\ \cdot \end{array}$$

We'll multiply each digit on the top row by 5

Write a decimal point on a new line

7 When we multiply by this 5, we're actually multiplying by 50, which is 10 times 5. So, we put a 0 in the first column on the new line as a place holder.

$$\begin{array}{r} 6 \cdot 3 \\ \times \quad 5 \quad 2 \\ \hline 1 \quad 2 \cdot 6 \\ \cdot 0 \end{array}$$

This 5 means 5 tens or 50

Put a 0 on a new line as a place holder

8 Now multiply 3 by 5. The answer is 15. Write the 5 in the column to the left of the decimal point. Carry the 1 into the next column to add on at the next stage.

$$\begin{array}{r} {}_1 \quad 6 \cdot 3 \\ \times \quad 5 \quad 2 \\ \hline 1 \quad 2 \cdot 6 \\ 5 \cdot 0 \end{array}$$

1 carried into the next column

Multiply 3 by 5

9 Next, multiply 6 by 5. The answer is 30. Add the 1 ten carried over from the previous step to make 31. Write 1 in the next available column and the 3 in the next column to the left.

$$\begin{array}{r} {}_1 \quad 6 \cdot 3 \\ \times \quad 5 \quad 2 \\ \hline 1 \quad 2 \cdot 6 \\ 3 \quad 1 \quad 5 \cdot 0 \end{array}$$

Multiply 6 by 5

The 1 carried is added to the number put in this column

10 Now we have multiplied each of the digits on the top row by all of the digits on the bottom row, we add the two lines in our answer together: 12.6 + 315.0 = 327.6

$$\begin{array}{r} {}_1 \quad 6 \cdot 3 \\ \times \quad 5 \quad 2 \\ \hline 1 \quad 2 \cdot 6 \\ + \quad 3 \quad 1 \quad 5 \cdot 0 \\ \hline 3 \quad 2 \quad 7 \cdot 6 \end{array}$$

Add the two lines in the answer together

11 So, 6.3 × 52 = 327.6

$$6.3 \times 52 = 327.6$$

The lattice method

There are several ways to do multiplication calculations, as you have seen. The lattice method, shown here, is very similar to long multiplication, but we write the numbers out in a grid instead of columns. We can use the lattice method for large whole numbers, and numbers with decimals.

The lattice method can be used for whole numbers and decimals.

1 Let's multiply 78 by 64 using the lattice method.

$$78 \times 64 = ?$$

2 The numbers in our calculation are both two digits long, so we draw a grid, or lattice, that is two boxes long and two boxes tall. Write the numbers in the calculation along the edges of the lattice.

3 Now, draw a diagonal line through each box from the top right to the bottom left. The numbers that we are going to write along each diagonal will have the same place value.

Extend the diagonal beyond the edge of the lattice

4 Next, multiply the digit at the top of each column by the digit at the end of each row. When we multiply 7 by 6, the answer is 42. Write 4 in the top of the box and 2 in the bottom of the box. We are separating the product into its tens and ones.

Write the tens above the line

Write the ones below the line

5 Continue multiplying the numbers at the top of each column and the end of each row until all the boxes are filled.

Write the product in each box

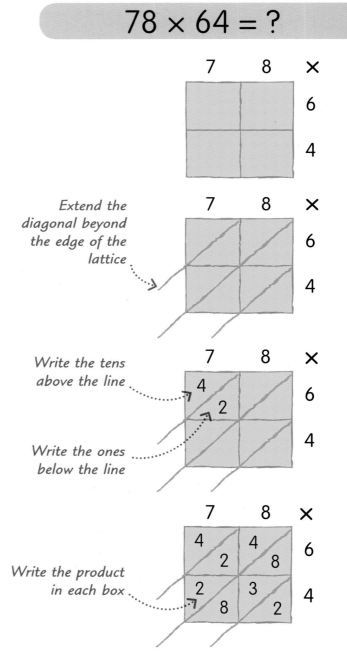

6 Starting from the bottom right corner, add the numbers along each diagonal. The first diagonal has just the number 2, so we write 2 at the edge end of the diagonal.

Write the total of the diagonal at the edge

7 Now add the numbers in the second diagonal: 8 + 3 + 8 = 19. Write 9 at the end of the diagonal and carry the 1 ten into the next diagonal to add on at the next stage.

The carried 1 ten goes here

8 Keep adding the numbers across each diagonal, until we reach the top left corner. We are left with the numbers 4, 9, 9, and 2. So, the answer is 4992.

Read the answer from the top left to the bottom right

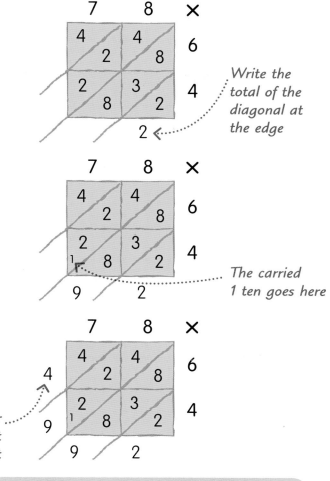

9 So, 78 × 64 = 4992

$$78 \times 64 = 4992$$

Multiplying decimals using the lattice method

We can use the lattice method to multiply decimals, too. We just need to find where the decimal points meet.

1 Let's multiply 3.59 by 2.8. First, write the two numbers along the edges of the lattice, including the decimal points. Work through the steps in the same way that we did with the whole numbers above.

2 Next, look down from the decimal point at the top and along from the decimal point at the side and find where they meet inside the lattice.

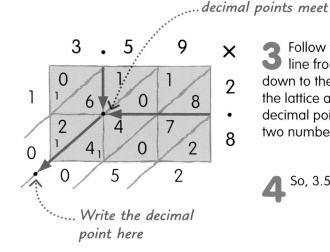

Find where the decimal points meet

Write the decimal point here

3 Follow the diagonal line from this point down to the bottom of the lattice and write the decimal point between the two numbers at the end.

4 So, 3.59 × 2.8 = 10.052

Division

Division is splitting a number into equal parts, or finding out how many times one number fits into another number. It doesn't always work out exactly. Sometimes there's a bit left over.

Division is sharing something out equally.

Division is sharing

When we divide something, like a number of apples, we share it out equally. Each part of a division calculation has its own special name.

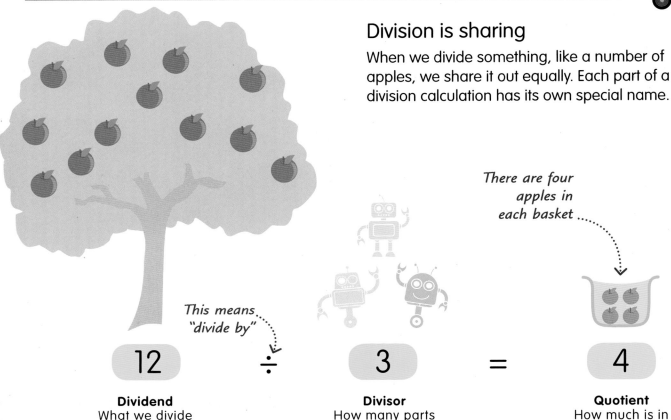

There are four apples in each basket

This means "divide by"

12	÷	3	=	4
Dividend What we divide		**Divisor** How many parts we divide it into		**Quotient** How much is in each part

1 Three robots have come to pick the 12 ripe apples on this tree. How many will each robot get? We need to divide!

2 If we divide, or share out, the 12 apples equally between the 3 robots, each robot gets 4 apples. So, 12 ÷ 3 = 4

One more apple

What happens if there are 13 apples, rather than 12? The 3 robots still get 4 apples each, but now there's 1 left over. We call the extra apple the remainder, and we put an "r" in front of it.

13	÷	3	=	4 r1

Division is the opposite of multiplication

If we know a multiplication fact, we can use it to find a division fact. This is because division is the opposite, or inverse, of multiplication. We can show this with our robots and apples.

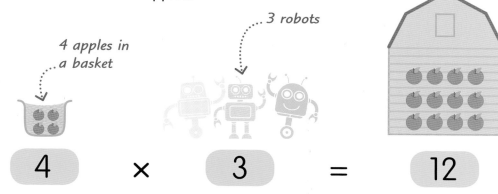

12 apples in the barn ......

... 3 robots

4 apples in ... a basket

1 The 3 robots are storing their apples. Each robot takes a basket of 4 apples and empties it into the barn. The total number of apples in the barn is 12, because 4 multiplied by 3 is 12.

$$4 \times 3 = 12$$

2 The multiplication to store the apples $(4 \times 3 = 12)$ is the inverse of the division we did to share them out $(12 \div 3 = 4)$. The 3 stays where it is, but the other numbers change places. So, if you know the multiplication, you just rearrange the numbers to find the division, and vice versa.

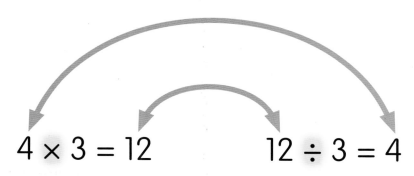

$$4 \times 3 = 12 \qquad 12 \div 3 = 4$$

Division is repeated subtraction

Division is also like taking away one number from another number again and again. We call this repeated subtraction. Let's see what happens when our robots start removing their apples from the barn.

Repeated subtraction is the inverse of repeated addition, which we looked at on page 99.

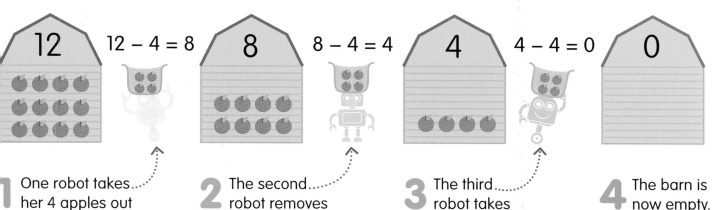

12    $12 - 4 = 8$ 8 $8 - 4 = 4$ 4 $4 - 4 = 0$ 0

1 One robot takes her 4 apples out of the barn. There are 8 apples left.

2 The second robot removes his 4 apples, leaving 4 apples in the barn.

3 The third robot takes the last 4 apples out of the barn.

4 The barn is now empty. This shows us that $12 \div 3 = 4$

Dividing with multiples

We've already used number lines to add, subract, and multiply. We can also use them to see how many times one number (the divisor) fits into another (the dividend). The division is easier if you jump forward in multiples of the divisor.

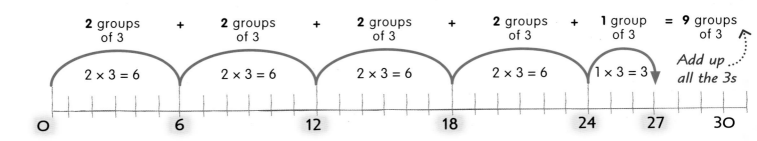

1 Let's calculate 27 ÷ 3. We'll start at 0 and make jumps of 2 groups of 3 each time. Each jump moves us 6 places.

$$27 \div 3 = ?$$

2 Four jumps gets us to 24. A last jump of 3 takes us to 27. We've jumped 9 groups of 3 in total, so that's the answer.

$$27 \div 3 = 9$$

3 If we made bigger jumps, we could get to the answer with fewer steps.

What about remainders?

Sometimes our jumps don't quite reach the target. In cases such as this, we're left with a remainder. Let's see what happens when we use a number line to divide 44 by 3.

> The bigger the multiples, the fewer steps you need.

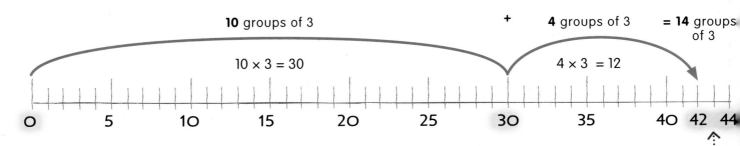

1 A first big jump of 10 groups of 3 moves us 30 places. Then a jump of 4 groups of 3 moves us on another 12.

$$44 \div 3 = ?$$

2 Our two jumps have taken us to 42, but we're 2 places short of 44. So our remainder is 2.

$$44 \div 3 = 14\ r2$$

We can't add another group of 3 without going past our target, so our remainder is 2

The division grid

We can take the multiplication grid (see page 106) and use it as a division grid. The numbers in the middle are the dividends – the numbers we want to divide. Those along the top and down one side are the divisors and the quotients.

1 Let's use our division grid to calculate 56 ÷ 7.

$$56 \div 7 = ?$$

2 First, we find the number we want to divide by. We go along the top blue row to 7.

3 Next, we move down the 7 column until we find the number we want to divide, which is 56.

4 Lastly, we move along the row from 56 until we reach 8 in the blue column on the left. This is the answer (quotient) to our division calculation.

×	1	2	3	4	5	6	7	8	9	10	11	12
1	1	2	3	4	5	6	7	8	9	10	11	12
2	2	4	6	8	10	12	14	16	18	20	22	24
3	3	6	9	12	15	18	21	24	27	30	33	36
4	4	8	12	16	20	24	28	32	36	40	44	48
5	5	10	15	20	25	30	35	40	45	50	55	60
6	6	12	18	24	30	36	42	48	54	60	66	72
7	7	14	21	28	35	42	49	56	63	70	77	84
8	8	16	24	32	40	48	56	64	72	80	88	96
9	9	18	27	36	45	54	63	72	81	90	99	108
10	10	20	30	40	50	60	70	80	90	100	110	120
11	11	22	33	44	55	66	77	88	99	110	121	132
12	12	24	36	48	60	72	84	96	108	120	132	144

5 So, 56 ÷ 7 = 8. This is the inverse of 7 × 8 = 56.

$$56 \div 7 = 8$$

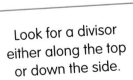
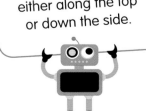

Look for a divisor either along the top or down the side.

TRY IT OUT

Grid lock!

Use the grid to find the answers to these division calculations.

Answers on page 319

1 A £72 competition prize is shared between 8 winners. How much does each person win?

2 A bag of 54 marbles is shared between 9 children. How many does each child get?

Division tables

We can list division facts in tables just like we list multiplication facts in multiplication tables. Division tables are the opposite, or inverse, of multiplication tables. You can use these tables to help you with division calculations.

1÷ table

1	÷	1	=	**1**
2	÷	1	=	**2**
3	÷	1	=	**3**
4	÷	1	=	**4**
5	÷	1	=	**5**
6	÷	1	=	**6**
7	÷	1	=	**7**
8	÷	1	=	**8**
9	÷	1	=	**9**
10	÷	1	=	**10**
11	÷	1	=	**11**
12	÷	1	=	**12**

2÷ table

2	÷	2	=	**1**
4	÷	2	=	**2**
6	÷	2	=	**3**
8	÷	2	=	**4**
10	÷	2	=	**5**
12	÷	2	=	**6**
14	÷	2	=	**7**
16	÷	2	=	**8**
18	÷	2	=	**9**
20	÷	2	=	**10**
22	÷	2	=	**11**
24	÷	2	=	**12**

3÷ table

3	÷	3	=	**1**
6	÷	3	=	**2**
9	÷	3	=	**3**
12	÷	3	=	**4**
15	÷	3	=	**5**
18	÷	3	=	**6**
21	÷	3	=	**7**
24	÷	3	=	**8**
27	÷	3	=	**9**
30	÷	3	=	**10**
33	÷	3	=	**11**
36	÷	3	=	**12**

4÷ table

4	÷	4	=	**1**
8	÷	4	=	**2**
12	÷	4	=	**3**
16	÷	4	=	**4**
20	÷	4	=	**5**
24	÷	4	=	**6**
28	÷	4	=	**7**
32	÷	4	=	**8**
36	÷	4	=	**9**
40	÷	4	=	**10**
44	÷	4	=	**11**
48	÷	4	=	**12**

5÷ table

5	÷	5	=	**1**
10	÷	5	=	**2**
15	÷	5	=	**3**
20	÷	5	=	**4**
25	÷	5	=	**5**
30	÷	5	=	**6**
35	÷	5	=	**7**
40	÷	5	=	**8**
45	÷	5	=	**9**
50	÷	5	=	**10**
55	÷	5	=	**11**
60	÷	5	=	**12**

6÷ table

6	÷	6	=	**1**
12	÷	6	=	**2**
18	÷	6	=	**3**
24	÷	6	=	**4**
30	÷	6	=	**5**
36	÷	6	=	**6**
42	÷	6	=	**7**
48	÷	6	=	**8**
54	÷	6	=	**9**
60	÷	6	=	**10**
66	÷	6	=	**11**
72	÷	6	=	**12**

Answers on page 319

TRY IT OUT

Tea-party teaser

Use the division tables to help you answer these tricky questions.

Imagine you have made 24 sandwiches for a tea-party. How many sandwiches will each person get if there are:

1 2 guests?

2 3 guests?

3 4 guests?

4 6 guests?

5 8 guests?

6 12 guests?

7 ÷ table

7	÷	7	=	1
14	÷	7	=	2
21	÷	7	=	3
28	÷	7	=	4
35	÷	7	=	5
42	÷	7	=	6
49	÷	7	=	7
56	÷	7	=	8
63	÷	7	=	9
70	÷	7	=	10
77	÷	7	=	11
84	÷	7	=	12

8 ÷ table

8	÷	8	=	1
16	÷	8	=	2
24	÷	8	=	3
32	÷	8	=	4
40	÷	8	=	5
48	÷	8	=	6
56	÷	8	=	7
64	÷	8	=	8
72	÷	8	=	9
80	÷	8	=	10
88	÷	8	=	11
96	÷	8	=	12

9 ÷ table

9	÷	9	=	1
18	÷	9	=	2
27	÷	9	=	3
36	÷	9	=	4
45	÷	9	=	5
54	÷	9	=	6
63	÷	9	=	7
72	÷	9	=	8
81	÷	9	=	9
90	÷	9	=	10
99	÷	9	=	11
108	÷	9	=	12

10 ÷ table

10	÷	10	=	1
20	÷	10	=	2
30	÷	10	=	3
40	÷	10	=	4
50	÷	10	=	5
60	÷	10	=	6
70	÷	10	=	7
80	÷	10	=	8
90	÷	10	=	9
100	÷	10	=	10
110	÷	10	=	11
120	÷	10	=	12

11 ÷ table

11	÷	11	=	1
22	÷	11	=	2
33	÷	11	=	3
44	÷	11	=	4
55	÷	11	=	5
66	÷	11	=	6
77	÷	11	=	7
88	÷	11	=	8
99	÷	11	=	9
110	÷	11	=	10
121	÷	11	=	11
132	÷	11	=	12

12 ÷ table

12	÷	12	=	1
24	÷	12	=	2
36	÷	12	=	3
48	÷	12	=	4
60	÷	12	=	5
72	÷	12	=	6
84	÷	12	=	7
96	÷	12	=	8
108	÷	12	=	9
120	÷	12	=	10
132	÷	12	=	11
144	÷	12	=	12

Dividing with factor pairs

You'll remember that a factor pair is two numbers that we multiply together to get another number (see pages 28 and 101). Factor pairs are just as useful in division as they are in multiplication.

FACTOR PAIRS OF 12

This is the multiplier

$1 \times 12 = 12$

$2 \times 6 = 12$

$3 \times 4 = 12$

$4 \times 3 = 12$

$6 \times 2 = 12$

$12 \times 1 = 12$

DIVISION FACTS OF 12

The multiplier of each factor pair is now the divisor

$12 \div 12 = 1$

$12 \div 6 = 2$

$12 \div 4 = 3$

$12 \div 3 = 4$

$12 \div 2 = 6$

$12 \div 1 = 12$

1 These are all the factor pairs of 12. The inverse of each multiplication fact is a division fact of 12. The multiplier of the factor pair becomes the divisor in the division fact.

2 If we divide 12 by one of the numbers from a factor pair, then the answer will be the other number in the pair. For example, $12 \div 3$ must be 4, because 3 and 4 are a factor pair of 12.

Factor pairs and multiples of 10

You can also use factor pairs when you are dividing with numbers that are multiples of 10. The only thing that's different is the zeros – all the other digits are the same. Here are some examples.

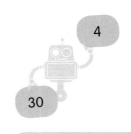 

$120 \div 30 = ?$

$120 \div 30 = 4$

$120 \div 60 = ?$

$120 \div 60 = 2$

$150 \div 50 = ?$

$150 \div 50 = 3$

1 Let's look at $120 \div 30$. The answer is 4. You know that 3 and 4 are a factor pair of 12, so 30 and 4 must be a factor pair of 120.

2 What about $120 \div 60$? Since 6 and 2 are a factor pair of 12, 60 and 2 must be a factor pair of 120. So the answer is 2.

3 This is also true of other multiples of 10. For example, 5 and 3 are a factor pair of 15, because $5 \times 3 = 15$. So the answer to $150 \div 50$ must be 3.

Checking for divisibility

A simple calculation or an observation about a number will often tell you whether or not it can be divided exactly (without a remainder) by a whole number. The checks in the table below will help you with your division.

A number is divisible by	If ...	Examples
2	If the last digit is an even number	**8, 12, 56, 134, 5000** are all divisible by 2
3	If the sum of all its digits is divisible by 3	**18** $1 + 8 = 9$ $(9 \div 3 = 3)$
4	If the number formed by the last two digits is divisible by 4	**732** $32 \div 4 = 8$ (we can divide 32 by 4 without a remainder, so 732 is divisible by 4)
5	If the last digit is 0 or 5	**10, 25, 90, 835, 1260** are all divisible by 5
6	If the number is even and the sum of all its digits is divisible by 3	**3426** $3 + 4 + 2 + 6 = 15$ $(15 \div 3 = 5)$
8	If the number formed by the last three digits is divisible by 8	**75 160** $160 \div 8 = 20$ (we can divide 160 by 8 without a remainder, so 75 160 is divisible by 8)
9	If the sum of the digits is divisible by 9	**6831** $6 + 8 + 3 + 1 = 18$ $(18 \div 9 = 2)$
10	If the last digit is 0	**10, 30, 150, 490, 10 000** are all divisible by 10
12	If the number is divisible by 3 and 4	**156** $156 \div 3 = 52$ and $156 \div 4 = 39$ (since 156 is divisible by 3 and 4, it's also divisible by 12)

Dividing by 10, 100, and 1000

Dividing by 10 is simple: you just shift the digits one place to the right on a place-value grid. By shifting the digits further to the right, you can also divide by 100 and 1000.

We can divide a number by 10, 100, or 1000 just by changing the place value of its digits.

1 Dividing by 10
To test this method, let's divide 6452 by 10. When we divide by 10, each digit becomes 10 times smaller. To show this, we move each digit one place to the right. This shows that 6452 ÷ 10 = 645.2

Th	H	T	O	$\frac{1}{10}$
6	4	5	2 .	
	6	4	5 .	2

Each digit shifts one place to the right

2 Dividing by 100
Now let's try dividing 6452 by 100. When we divide by 100, each digit becomes 100 times smaller. To show this, we move each digit two places to the right. So, 6452 ÷ 100 = 64.52

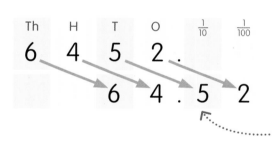

Th	H	T	O	$\frac{1}{10}$	$\frac{1}{100}$
6	4	5	2 .		
		6	4 .	5	2

Each digit shifts two places to the right

3 Dividing by 1000
Lastly, we'll divide 6452 by 1000. When we divide by 1000, each digit becomes 1000 times smaller. To show this, we move each digit three places to the right. This means 6452 ÷ 1000 = 6.452

Th	H	T	O	$\frac{1}{10}$	$\frac{1}{100}$	$\frac{1}{1000}$
6	4	5	2 .			
			6 .	4	5	2

Each digit shifts three places right

TRY IT OUT

Factory work

Can you use the "shift to the right" method to find the answers to these questions?

Answers on page 319

1 A factory owner shares £182 540 among 1000 workers. How much does each worker get?

2 The factory made 455 700 cars this year. That's 100 times more cars than it made 50 years ago. How many cars did it make then?

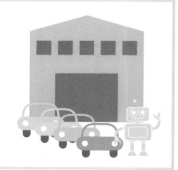

Dividing by multiples of 10

If your divisor (the number you're dividing by) is a multiple of 10, you can split the calculation into two easier steps. For example, instead of dividing by 50, you divide first by 10 and then by 5.

> To split up a multiple of 10 for this kind of division, break the multiple into 10 and its other factor.

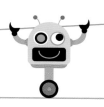

1 This calculation asks how many times 30 fits into 6900. Although we're dividing a big number, it's not as difficult as it looks.

$$6900 \div 30 = ?$$

2 Since 30 is a multiple of 10, we can split the division. Dividing in stages by 10 and 3 is easier than dividing by 30 all at once.

$$6900 \div 10 \div 3$$

.....*Stage one*   .....*Stage two*

3 First, we divide 6900 by 10. See page 136 (opposite) if you need help with this. The answer is 690.

$$6900 \div 10 = 690$$

4 Next, we divide 690 by 3. The answer is 230.

$$690 \div 3 = 230$$

5 So, 6900 ÷ 30 = 230

$$6900 \div 30 = 230$$

TRY IT OUT

Mind-boggling multiples

The divisors in these questions are multiples of 10. Split up the multiples, then find the answers.

Answers on page 319

1 A class of 20 children has to deliver 860 leaflets to advertise the school craft fair. If they share the work equally, how many leaflets should each child take?

2 The children also make some bead bracelets to sell at the fair. Each bracelet contains 40 beads. How many bracelets do they make with 1800 beads?

Partitioning for division

When you're dividing a number with two or more digits, it helps to break that number down, or partition it, into smaller numbers that are easier to work with.

How to partition

The first step in partitioning for division is to break the number we're dividing (the dividend) into two smaller numbers. It's often a good idea to break the dividend into a multiple of 10 and another number. Then we divide each of these two numbers by the number we're dividing by (the divisor). Finally, we add our two answers (or quotients) to get the final answer.

Break up 147 into easy-to-divide parts

140

7

1 Let's divide 147 by 7 using partitioning.

$$147 \div 7 = ?$$

2 We're going to partition 147 into 140 and 7.

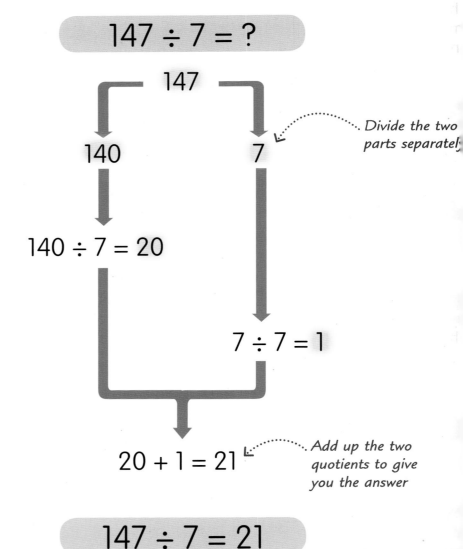

147

140   7

Divide the two parts separately

3 First, we divide 140 by 7. We know from the multiplication table for 7 that $7 \times 10 = 70$, so $7 \times 20 = 140$. This tells us that $140 \div 7 = 20$

$$140 \div 7 = 20$$

4 Now we divide 7 by 7. That's easy! The answer is 1.

$$7 \div 7 = 1$$

5 Now we simply add up the answers we got from dividing the parts separately: $20 + 1 = 21$

$$20 + 1 = 21$$

Add up the two quotients to give you the answer

6 So, $147 \div 7 = 21$

$$147 \div 7 = 21$$

Including remainders

Sometimes, dividing by partitioning leaves us with remainders. But the method we've just seen still works – we simply have to include the remainders when we add up our answers (or quotients) at the end.

1 Imagine you're going on holiday in 291 days and you want to know how many weeks you have to wait until the holiday begins. You know there are 7 days in a week, so you need to divide 291 by 7 to find out the number of weeks.

2 Since we know from the multiplication table for 7 that $7 \times 4 = 28$, we also know that $7 \times 40 = 280$, which is very close to, but not more than, the dividend (291). Let's partition 291 into 280 and 11.

3 As we know that $7 \times 40 = 280$, we also know that $280 \div 7 = 40$

4 Now we divide 11 by 7. The answer is 1 remainder 4.

5 Adding up our quotients and including the remainder gives the final answer 41 r4.

6 So, $291 \div 7 = 41$ r4

7 Remember, we're counting in weeks, so we can also write the answer as 41 weeks and 4 days.

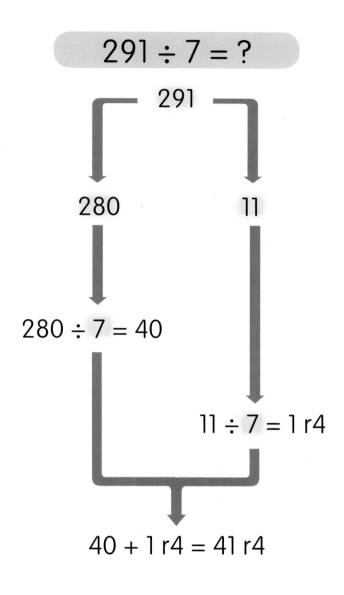

$$291 \div 7 = ?$$

291

280   11

$$280 \div 7 = 40$$

$$11 \div 7 = 1 \text{ r4}$$

$$40 + 1 \text{ r4} = 41 \text{ r4}$$

$$291 \div 7 = 41 \text{ r4}$$

Expanded short division

Short division is a method we use when the number we are dividing by (the divisor) has only one digit. To make the calculation easier, we use expanded short division. In this method, we subtract multiples, or "chunks", of the divisor.

1 To try out expanded short division, let's divide 156 by 7.

$$156 \div 7 = ?$$

2 First, we write the number we want to divide (the dividend). In this case, it's 156. We draw a division bracket (like an upturned "L") around it. We put the divisor, 7, outside the bracket, to the left of 156.

H   T   O

$$7 \overline{\smash{\big)}\ 1 \quad 5 \quad 6}$$

You may find it helps to label the place values

Division bracket

3 Now we're ready to begin dividing. Expanded short division is just like repeated subtraction, but instead of taking away 7 repeatedly, we subtract much bigger chunks of the number each time. To start, we'll take away 70, which is 10 groups of 7. So, we subtract 70 from 156, which leaves 86.

H   T   O

$$
\begin{array}{r}
7\,\overline{\smash{\big)}\ \cancel{1}\ {}^{1}5\ \ 6} \\
-\ 7\ \ 0 \quad (7 \times 10) \\
\hline
8\ \ 6
\end{array}
$$

We write down how many 7s we've taken away

Draw a line and write what's left over here, making sure you keep the place values lined

4 We have 86 left over, so we can subtract another chunk of 70 from it. That leaves 16. We've now subtracted 20 groups of 7 from 156.

H   T   O

$$
\begin{array}{r}
7\,\overline{\smash{\big)}\ 1\ \ 5\ \ 6} \\
-\ 7\ \ 0 \quad (7 \times 10) \\
\hline
8\ \ 6 \\
-\ 7\ \ 0 \quad (7 \times 10) \\
\hline
1\ \ 6
\end{array}
$$

86 − 70 = 16

Record another ten groups of 7

Expanded short division uses repeated subtraction, which we looked at on page 129.

5 Now we've only got 16 left from our original dividend of 156. That number's too small to subtract another 70, so we need to find the largest number of 7s we can take away from 16. The answer's 2 of course, since 7 × 2 = 14

6 Next, we take away 14 from 16. That leaves us with 2. We can't take any more 7s away from 2, so we've come to the end of our subtractions. The left-over 2 is the remainder.

```
        H   T   O
    7 | 1   5   6
    −       7   0    (7 × 10)
    _____
            8   6
    −       7   0    (7 × 10)
        _____
            1   6
    −       1   4    (7 × 2)
        _____
                2
```

This is the remainder ....

Keep writing down the number of 7s

7 The last step is to add up how many 7s we've taken away. That's why we wrote them down beside our calculation as we went along. So, 10 + 10 + 2 = 22 groups of 7. Write 22 above the bracket, then put "r2" beside it to show that 7 doesn't go into 156 exactly.

Put the total number of 7s here

```
        H   T   O
            2   2   r2
    7 | 1   5   6
    −       7   0    (7 × 10)
    _____
            8   6
    −       7   0    (7 × 10)
        _____
            1   6
    −       1   4    (7 × 2)
        _____
                2      22
```

8 So, 156 ÷ 7 = 22 r2

156 ÷ 7 = 22 r2

Add up how many 7s we've subtracted ....

TRY IT OUT

Expand your skills

Try using expanded short division to do these division calculations.

Answers on page 319

1 196 ÷ 6 = ?
Start by subtracting 30 groups of 6.

2 234 ÷ 5 = ?

If you work with bigger chunks, you'll be able to do the division with fewer subtractions.

Short division

Short division is another method for working out division
calculations on paper when the divisor is a single-digit number.
Compared with expanded short division (see pages 140-41), you
have to do more calculation in your head and less writing down.

1 Let's divide 156 by 7 using
short division.

$$156 \div 7 = ?$$

2 Write out the calculation
like this.

H	T	O

7) 1 5 6

*Use place value
columns if it
helps you*

3 Now we're going to divide
each of the digits in the
dividend, 156, by 7. We'll start
with the first digit, which is 1.

H	T	O

7) 1 5 6

*Begin by
dividing the
first digit
of 156 by 7*

4 Since 1 can't be divided by 7, we
write nothing over the 1 above the
division bracket. We carry over this 1
into the tens column. This carried over
1 stands for 1 hundred, which is the
same as 10 tens.

H	T	O

7) 1 5 6

*Carry the
1 hundred over
to the tens
column*

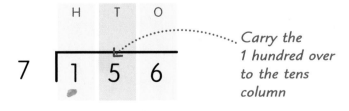

Remember, the number
being divided is the
dividend, and the number it
is divided by is the divisor.

5 Because we carried over the 1 from the hundreds column, we don't divide 5 by 7, instead we divide 15 by 7. We know that 7 × 2 = 14, so there are two 7s in 15 with 1 left over. Write the 2 above the division bracket in the tens column, and carry over the remaining 1 to the ones column. This 1 stands for 1 ten, or 10 ones.

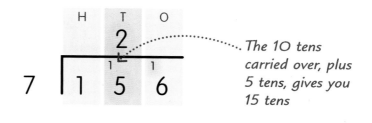

The 10 tens carried over, plus 5 tens, gives you 15 tens

6 Now look at the ones column. Because we carried over the 1 from the tens column, we divide 16 by 7. There are two 7s in 16 with 2 left over. Write the 2 above the division bracket in the ones column, and write the remainder next to it.

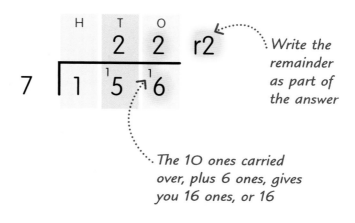

Write the remainder as part of the answer

The 10 ones carried over, plus 6 ones, gives you 16 ones, or 16

7 So, 156 ÷ 7 = 22 r2.

$$156 \div 7 = 22 \text{ r2}$$

TRY IT OUT

Test your skills

Glob has been busy sorting out screws into piles of different colours. Now she needs to divide each pile into groups, ready for use. Can you use short division to help her work out how many groups she can make with each pile?

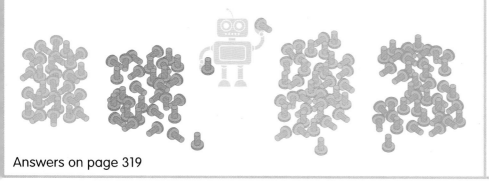

Answers on page 319

1 In the pink group, there are 279 screws, and Glob needs to divide these into groups of 9.

2 There are 286 blue screws, and she needs groups of 4.

3 There are 584 yellow screws, and she needs groups of 6.

4 There are 193 green screws, and she needs groups of 7.

Expanded long division

When the number we are dividing by (the divisor) has more than one digit, we use a method of calculation called long division. Here, we look at expanded long division. There's also a shorter version just called long division (see pages 146-47).

1 To see what expanded long division is like, we'll divide 4728 by 34.

$$4728 \div 34 = ?$$

2 Before we begin dividing, we write down the number we want to divide, the dividend, which is 4728. Then we draw a division bracket around it. We put the divisor, 34, outside the bracket, to the left of 4728.

	Th	H	T	O
34	4	7	2	8

You may find it useful to label the columns to show place values

3 Now we're all set to start dividing. Just as we did with expanded short division, we'll take away big chunks of the number each time. The easiest big chunk to take away is 100 groups of 34, which is 3400. When we subtract 3400 from 4728, we're left with 1328. We write the number of 34s on the right.

	Th	H	T	O	
34	4	7	2	8	
−	3	4	0	0	(34 × 100)
	1	3	2	8	

We write down how many 34s we subtracted

Draw a line and write what's left over here, keeping digits with the same place values lined up

4 We can't subtract another 3400 from 1328, so we'll need to use a smaller chunk. Fifty groups of 34 would be 1700. Forty groups would be 1360. Both numbers are too large. What about 30 groups of 34? That gives us 1020. Let's subtract 1020 from 1328, which leaves us with 308.

	Th	H	T	O	
34	4	7	2	8	
−	3	4	0	0	(34 × 100)
	1	3	2	8	
−	1	0	2	0	(34 × 30)
		3	0	8	

1328 − 1020 = 308

Record another 30 groups of 34

5 We've got 308 left from our original dividend of 4728. That's not quite enough to take away a chunk of 10 34s, which would be 340. But we can subtract nine 34s, which is 306.

6 When we take away 306 from 308, we're left with 2. We can't take any more 34s away, so that's the end of our subtractions. The 2 is our remainder.

There is a remainder of 2

	Th	H	T	O	
34	4	7	2	8	
−	3	4	0	0	(34 × 100)
	1	3	2	8	
−	1	0	2	0	(34 × 30)
		3	0	8	
−		3	0	6	(34 × 9)
				2	

Keep writing down the number of 34s we've subtracted

7 Finally, let's add up how many 34s we took away, which we listed beside our calculation as we went along. So, 100 + 30 + 9 = 139 groups of 34. Write 139 above the bracket, then put "r2" beside it to show that 34 goes into 4728 139 times with a remainder of 2.

	Th	H	T	O	
		1	3	9	r2
34	4	7	2	8	
−	3	4	0	0	(34 × 100)
	1	3	2	8	
−	1	0	2	0	(34 × 30)
		3	0	8	
−		3	0	6	(34 × 9) +
				2	139

Write the total number of 34s here

Add up how many 34s we've subtracted

The bigger the chunks you work with, the fewer subtractions there are.

8 So, 4728 ÷ 34 = 139 r2

4728 ÷ 34 = 139 r2

TRY IT OUT

A fishy problem!

A fisherman catches 6495 fish. He sells them to 43 fish shops, giving each shop the same amount. Any fish left over he gives to his cats.

Answers on page 319

1 Can you use expanded long division to work out how many fish each shop gets?

2 How many are left for the cats?

Long division

In expanded long division (see pages 144-45), we divide by subtracting multiples of the divisor in chunks. Long division is a different method, in which we divide each digit of the number we're dividing (the dividend) in turn.

1 To see how long division works, we'll divide 4728 by 34.

$$4728 \div 34 = ?$$

2 We start by writing the number we want to divide, which is 4728. Then we draw a division bracket around it. We put the divisor, 34, outside the bracket, immediately to the left of 4728.

Th	H	T	O
34	4	7	2	8

You may find it useful to label the columns to show the place values

3 Now we try to divide the first digit of the dividend by 34. 34 won't go into 4, so we look to the next digit and divide 47 by 34. The answer is 1. Write 1 above the bracket, over the 7. Write 34 beneath 47. Subtract 34 from 47 to find the remainder, which is 13. Write this in at the bottom.

	Th	H	T	O
		1		
34	4	7	2	8
−		3	4	
		1	3	

Write down how many 34s go into 47 here

Draw a line and write the total of the subtraction beneath it

4 We now bring down the next digit in the dividend to sit next to the 13 we just wrote down, to change the number 13 into 132.

	Th	H	T	O
		1		
34	4	7	2	8
−		3	4	
		1	3	2

When bringing down the next digit, keep it in its place-value column

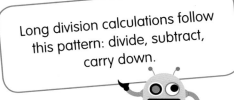

Long division calculations follow this pattern: divide, subtract, carry down.

5 Now divide 132 by 34. Let's split 34 into tens and ones (30 and 4) to make this easier. We know that 30 × 3 is 90, and 4 × 3 is 12, so 3 × 34 = 102. Write a 3 on the bracket above the 2. Write 102 beneath 132. Subtract 102 from 132 to find the remainder, which is 30.

6 Once again, bring down the next digit in the dividend to sit next to the 30 we just wrote down, to change the number 30 into 308.

7 Now divide 308 by 34. We know that 3 × 9 = 27, so 30 × 9 must be 270. We also know 9 × 4 = 36. And 270 + 36 = 306. So, 9 × 34 is 306. Write the 9 above the bracket, over the 8. This represents 9 × 34. Write 306 beneath 308, then subtract 306 from 308. The remainder is 2. Write the remainder into the answer on the bracket.

8 So, 4728 ÷ 34 = 139 r2

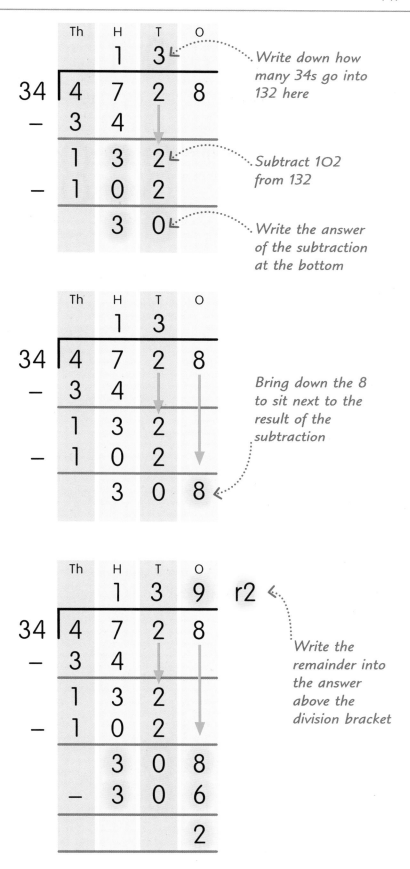

Write down how many 34s go into 132 here

Subtract 102 from 132

Write the answer of the subtraction at the bottom

Bring down the 8 to sit next to the result of the subtraction

Write the remainder into the answer above the division bracket

$$4728 \div 34 = 139 \text{ r2}$$

Converting remainders

We can convert the remainder in the answer to a division calculation into either a decimal or a fraction.

When you write your answer above the division bracket, line up the decimal point with the decimal point below the bracket.

Converting remainders into decimals

If the answer to a division calculation has a remainder, we can convert that into a decimal by simply adding a decimal point to the dividend and continuing with the calculation.

1 Let's divide 75 by 6 using expanded short division and convert the remainder into a decimal.

$$75 \div 6 = ?$$

2 Start by writing out the calculation like this.

Label the columns to show place values

3 First, divide the first digit in the dividend, 7, by 6. As 6 can go into 7 only once, write 1 above the 7 on the division bracket, in the tens column. Write the 6 beneath the 7, then subtract this 6 from 7 to get your remainder, which is 1.

Write down how many 6s you've subtracted from 7

Draw a line and write what's left over here, keeping the place values lined up

4 Now we move on to the second digit in the dividend which is 5. Bring this down to sit next to the 1 at the bottom of the calculation. Divide 15 by 6. We know 6 × 2 = 12, so write 2 on the division bracket in the ones column. Write 12 beneath 15 and subtract 12 from 15. The answer is 3. This is the remainder.

Bring down the 5

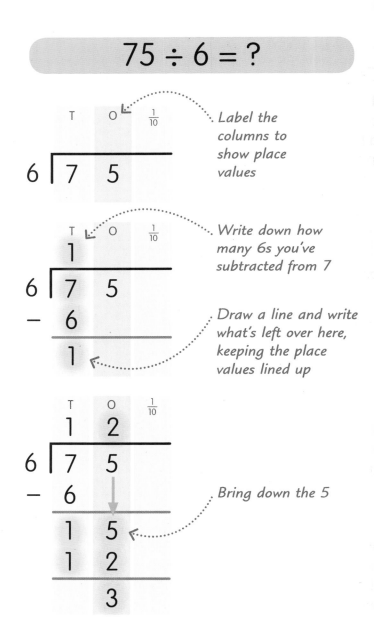

5 To turn this remainder 3 into a decimal, continue calculating. Place a decimal point at the end of the dividend and put a zero next to it. Add another decimal point above the division bracket, with a tenths column to the right. Bring down the new zero in the dividend to sit by the remainder 3. Now divide 30 by 6. We know that $6 \times 5 = 30$, so the answer is 5. Write this on the division bracket in the tenths column.

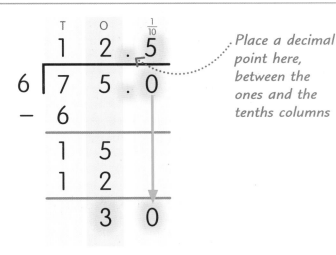

Place a decimal point here, between the ones and the tenths columns

6 As there's no remainder, we can end our calculation here. So, $75 \div 6 = 12.5$

$$75 \div 6 = 12.5$$

Converting remainders into fractions

It's simple to convert remainders into fractions. First, we carry out the division calculation. To turn the remainder into a fraction, we simply write the remainder as the numerator in the fraction and the divisor as the denominator.

The numerator is the top number in a fraction. The denominator is the one below.

1 Here, expanded short division has been used to divide 20 by 8. The answer is 2 r4.

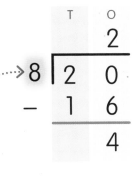

Use the divisor as the denominator in the fraction

Use the remainder as the numerator in the fraction

2 So, the remainder is 4/8. We know that 4/8 is the same as 2/4, which is the same as 1/2, so we can use the fraction 1/2 instead.

$$r4 = \frac{4}{8} = \frac{2}{4} = \frac{1}{2}$$

3 So, $20 \div 8 = 2\frac{1}{2}$. We can tell that our remainder is correct, because we know that half of 8 is 4, so a remainder of 4 can be written as 1/2.

$$20 \div 8 = 2\frac{1}{2}$$

Dividing with decimals

Dividing a number by a decimal number or dividing a decimal number is simple if you know how to divide whole numbers and how to multiply numbers by multiples of 10 (see pages 108-109).

Dividing by a decimal

When a divisor (the number you're dividing by) is a decimal number, first multiply it by 10 as many times as it takes to give you a whole number. You also have to multiply the dividend (the number being divided) by 10 the same number of times. Then do the division calculation and the answer will be the same as it would if you did the calculation without multiplying first.

> Multiply both the dividend and divisor by 10 until the decimal number you're working with becomes a whole number.

1 Let's divide 536 by 0.8

$$536 \div 0.8 = ?$$

2 First, multiply both the divisor and the dividend by 10. Then 536 becomes 5360 and, 0.8 becomes 8.

$$536 \times 10 = 5360$$

$$0.8 \times 10 = 8$$

3 Now carry out a division calculation. We can see from the completed calculation shown here that $5360 \div 8 = 670$

	Th	H	T	O
		6	7	0
8	5	3	6	0
−		4	8	
			5	6

.... *You'll need four place-value columns for this calculation*

4 So, the answer to both 536 ÷ 0.8 and 5360 ÷ 8 is 670.

$$536 \div 0.8 = 670 \text{ and } 5360 \div 8 = 670$$

Dividing a decimal

If it is the dividend (the number being divided) that is the decimal number, simply carry out the calculation as you would if there were no decimal point there. Make sure you write the decimal point into the answer in the correct place – directly above the one in the dividend.

1 Let's divide 1.24 by 4.

$$1.24 \div 4 = ?$$

2 Because the divisor (the number we are dividing by) is greater than the dividend, we know the answer will be less than 1. Write out the calculation with a division bracket. Now we can begin calculating.

O	$\frac{1}{10}$	$\frac{1}{100}$

$$4\ \overline{|\ 1\ .\ 2\ \ 4}$$

You'll need place value columns for decimal places

3 As 4 won't go into 1, write a zero on the division bracket above the 1 and a decimal point next to it. Now we look to the next digit in the dividend and divide 12 by 4. We know that 4 × 3 = 12, so we write the 3 on the bracket above the 2, after the decimal point. Write the 1.2 beneath the 1.2 in the dividend. Subtract 1.2 from 1.2, which gives us 0.

O	$\frac{1}{10}$	$\frac{1}{100}$

$$0\ .\ 3$$
$$4\ \overline{|\ 1\ .\ 2\ \ 4}$$
$$-\ \ 1\ .\ 2$$
$$\overline{0}$$

Keep the decimal points lined up, between the ones and tenths columns

4 Now carry down the final digit in the dividend, which is 4, to sit next to the 0 at the bottom of the calculation.

O	$\frac{1}{10}$	$\frac{1}{100}$

$$0\ .\ 3$$
$$4\ \overline{|\ 1\ .\ 2\ \ 4}$$
$$-\ \ 1\ .\ 2$$
$$\overline{0\ \ 4}$$

Bring down the 4 to the bottom of the calculation

5 Next, divide 4 by 4. The answer is 1. Write 1 on the division bracket above the 4 in the hundredths column. There's no remainder, so the calculation ends at this point.

O	$\frac{1}{10}$	$\frac{1}{100}$

$$0\ .\ 3\ \ 1$$
$$4\ \overline{|\ 1\ .\ 2\ \ 4}$$
$$-\ \ 1\ .\ 2$$
$$\overline{0\ \ 4}$$

Divide 4 by 4

6 So, 1.24 ÷ 4 = 0.31

$$1.24 \div 4 = 0.31$$

The order of operations

Some calculations are more complex than just two numbers with one operation. Sometimes we need to carry out calculations where there are several different operations to do. It's very important that we know which order to do them in so that we get the answer right.

BODMAS

We can remember the order that we should do calculations by learning the word "BODMAS" (or "BIDMAS"). It stands for brackets, orders (or indices), division, multiplication, addition, and subtraction. We should always work out calculations in this order, even if they are ordered differently when the calculation is written down.

$$4 \times \mathbf{(2 + 3)} = 20$$

1 Brackets
Look at this calculation. Two of the numbers are inside a pair of brackets. Brackets tell us that we must work out that part first. So, first we must find the sum of 2 + 3, then multiply 4 by that sum to find the total.

$$5 + 2 \times \mathbf{3^2} = 23$$

2 Orders (or indices)
Powers or square roots are known as orders or indices. We looked at these types of numbers on pages 36-39. We work these out after brackets. Here, we first work out 3^2 is 9, then $2 \times 9 = 18$, and finally add 5 to get 23.

$$6 + \mathbf{4 \div 2} = 8$$

3 Division
We work out division and multiplication calculations next. In this example, even though the division is written after the addition, we divide first. So, $4 \div 2 = 2$ and then $6 + 2 = 8$

$$\mathbf{8 \div 2} \times 3 = 12$$

4 Multiplication
Division and multiplication are of equal importance, so we work them out from left to right through a calculation. Look at this example. We divide first, then multiply: $8 \div 2 \times 3 = 4 \times 3 = 12$

$$9 \div 3 + \mathbf{12} = 15$$

5 Addition
Finally, we do any addition and subtraction calculations. Look at this calculation. We know that we do division before addition, so: $9 \div 3 + 12 = 3 + 12 = 15$

$$\mathbf{10 - 3} + 4 = 11$$

6 Subtraction
Addition and subtraction are of equal importance, like multiplication and division, so we work them out from left to right. Here, we subtract first, then add: $10 - 3 + 4 = 7 + 4 = 11$

Using BODMAS

If you can remember BODMAS, even calculations that look really tough are straightforward.

1 Let's give this tricky calculation a go.

$$17 - (4 + 6) \div 2 + 36 = ?$$

2 We know that we need to work out the brackets first, so we need to add 4 and 6, which equals 10. We can now write the calculation as: $17 - 10 \div 2 + 36$

$$17 - 10 \div 2 + 36 = ?$$

3 There are no orders in this calculation, so we divide next: $10 \div 2 = 5$. So, now we can write the calculation as: $17 - 5 + 36$

$$17 - 5 + 36 = ?$$

4 Now we can work from left to right and work out the addition and subtraction calculations one by one. Subtracting 5 from 17 gives 12. Finally, we add 36 to 12 to give 48.

$$12 + 36 = 48$$

5 So, $17 - (4 + 6) \div 2 + 36 = 48$

$$17 - (4 + 6) \div 2 + 36 = 48$$

TRY IT OUT

Follow the order

Now it's up to you. Use the order of operations and see if you can work out the correct answers to these calculations.

1 $12 + 16 \div 4 + (3 \times 7) = ?$

2 $4^2 - 5 - (12 \div 4) + 9 = ?$

3 $6 \times 9 + 13 - 22 \div 11 = ?$

Answers on page 319

BODMAS stands for:
Brackets
Orders
Division
Multiplication
Addition
Subtraction

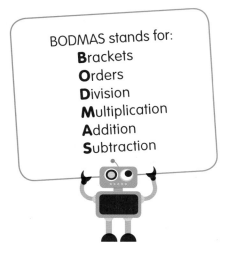

Arithmetic laws

Whenever we're calculating, it helps to remember three basic rules called the arithmetic laws. These are especially useful when we're working on a calculation with several parts.

The commutative law

When we add or multiply two numbers, it doesn't matter which order we do it in – the answer will be the same. This is called the commutative law.

$$+ \quad = \quad +$$

1 Addition
Look at these fish. Adding 6 to 5 gives 11 fish. Adding 5 to 6 also gives 11 fish. We can add numbers in any order and still get the same total.

$$5 + 6 = 11 \qquad 6 + 5 = 11$$

3 fish 2 times, or 3 × 2 ⋯⋯

2 fish 3 times, or 2 × 3 ⋯⋯

2 Multiplication
Here we have 3 fish 2 times, giving a total of 6 fish. If we have 2 fish 3 times, we also have a total of 6 fish. It doesn't matter what order we multiply the numbers, the product is the same.

$$3 \times 2 = 6 \qquad 2 \times 3 = 6$$

The associative law

When we add or multiply three or more numbers, the way we group the numbers doesn't affect the result. This is the associative law.

1 Addition
The associative law helps us to add together tricky numbers, like 136 + 47.

$$136 + 47$$

2 We can partition 47 into 40 + 7. If we work out this calculation, the answer is 183.

$$136 + (40 + 7) = 183$$

3 We can move the brackets to make the calculation simpler. Adding 136 and 40 first, then the 7, also gives 183.

$$(136 + 40) + 7 = 183$$

The distributive law

Multiplying a number by some numbers added together will give the same answer as multiplying each number separately. We call this the distributive law.

> When a calculation has numbers in brackets, work out the part in the brackets first. We looked at the order of operations on pages 152-53.

1 Let's see how the distributive law can help us to find 3×14.

$$3 \times 14 = ?$$

2 It's quite a hard calculation if we don't know our multiplication tables for 3 all the way to 14, so let's split 14 into 10 + 4, which is easier to work with.

$$3 \times (10 + 4) = ?$$

3 Next, we can make the calculation simpler to work out by distributing the number 3 to each of the numbers in the brackets.

$$(3 \times 10) + (3 \times 4) = ?$$

4 Now we can solve the two brackets before adding them together:
$(3 \times 10) + (3 \times 4) = 30 + 12 = 42$

$$30 + 12 = 42$$

5 So, by breaking 14 into simpler numbers and distributing the 3 between them, we've found that $3 \times 14 = 42$

$$3 \times 14 = 42$$

....

Multiplication

1 The associative law is also helpful when we need to multiply by a tricky number, like 6×15.

$$6 \times 15 = ?$$

2 We can break 15 into its factors 5 and 3. If we then work out this calculation, the answer is 90.

$$6 \times (5 \times 3) = 90$$

3 The associative law allows us to move the brackets to make it easier. If we find 6×5 before multiplying by 3, the answer is still 90.

$$(6 \times 5) \times 3 = 90$$

Using a calculator

A calculator is a machine that can help us to work out the answers to calculations. It's important that we know how to do calculations in our heads and with written methods, but sometimes using a calculator can make calculating quicker and easier.

Always double-check your answer when you are using a calculator, because it's easy to make a mistake by accidentally pressing the wrong keys.

Calculator keys

Most calculators have the same basic keys, just like this one. To use a calculator, we simply type in the calculation we want to work out, then press the [=] key. Let's take a look at what each key does.

The display shows the numbers that have been typed in or the answer

1 ON and CLEAR key
This is the key we press to turn the calculator on or to clear the display, taking the value displayed back to zero.

2 Number keys
The main part of the calculator's keypad are the numbers 0 to 9. We use these keys to enter the numbers in a calculation.

3 Decimal point key
We press this key if we are calculating with a decimal number. To enter 4.9, we press [4], then the decimal point [.], followed by [9].

4 Negative key
This key changes a positive number into a negative number, or a negative number into a positive number.

5 Arithmetic keys
All calculators have keys for adding [+], subtracting [−], multiplying [×], and dividing [÷]. If we wanted to calculate 14 × 27, we would press [1], [4], [×], [2], [7], then [=].

Calculator questions

Now that you know all of the important keys on the calculator and how to use them, see if you can work out the answers to these questions using a calculator.

Answers on page 319

1 983 + 528 = ?   4 39 × 64 = ?

2 7.61 – 4.92 = ?   5 697 ÷ 41 = ?

3 –53 + 21 = ?   6 40% of 600 = ?

6 Memory keys
Sometimes it can be useful to get a calculator to remember an answer, so that we can come back to it later. [M+] adds a number to the calculator's memory and [M–] removes that number. [MR] uses the number that is stored in the memory, without us needing to key it in and [MC] clears the memory.

7 Square root key
This key tells us the square root of a number. We use this in more advanced mathematics.

8 Percentage key
The [%] key can be used to work out percentages. It works a little differently on some calculators compared with others.

9 Equals key
This key is the "equals" key. When we have entered a calculation on the keypad, for example 14 × 27, we press [=] to reveal the answer on the calculator's display.

Estimating answers

When you use a calculator, it's easy to make mistakes by pressing the wrong keys. One way you can make sure your answer is right is to estimate what the answer should be. We looked at estimating on pages 24-25.

$$307 \times 49 = ?$$

1 Let's estimate the answer to 307 × 49

$$300 \times 50 = ?$$

2 It's quite tricky to work out in our heads so we can round the numbers up or down. Round 307 down to 300, and round 49 up to 50.

$$300 \times 50 = 15\,000$$

3 300 × 50 gives the answer 15 000, so the answer to 307 × 49 will be close to 15 000.

4 If we used the calculator to find 307 × 49 and got the answer 1813, then we would know it's incorrect and that we missed a number when keying it in. This is because estimating told us that the answer should be close to 15 000.

b×h

m^2

kg

°c

During history, people have used many different systems of measurement to describe the real world. But most countries now use the same system, called metric, to measure how big, heavy, or hot things are. It's easy to calculate with metric measurements. It's also simple to convert from one kind of metric measurement to another.

Length

Length is the distance between two points. We can measure distances in metric units called millimetres (mm), centimetres (cm), metres (m), and kilometres (km).

Metres and kilometres

We can use lots of different words to describe lengths, but they all mean the distance between two points.

1 Height means how far something is from the ground. But it's really no different from length, so we measure it in the same units. This tall building has a height of 700 m.

2 The width of something is a measure of how far it is from side to side. It's also a type of length. The width of this building is 250 m.

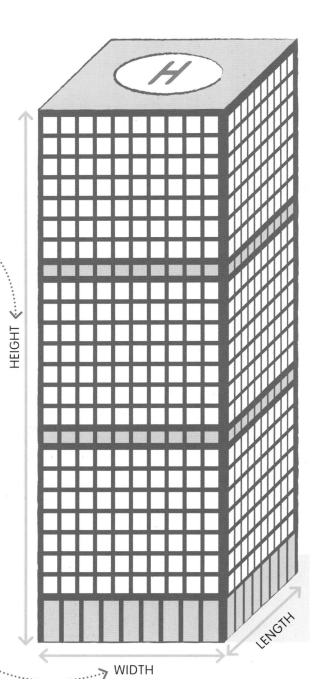

3 Another unit of length is the kilometre. There are 1000 m in 1 km. The helicopter is flying at a height of 1 km.

4 We can convert the height of the helicopter into metres by multiplying by 1000. So, the helicopter is 1000 m off the ground.

5 Another word we use for length is "distance", which means how far one place is from another. Long distances are measured in kilometres.

HEIGHT

LENGTH

WIDTH

Length, width, height, and distance are all measured using the same units.

Centimetres and millimetres

Metres and kilometres are great for measuring big things but less useful for measuring things that are much smaller. We can use units called centimetres and millimetres to measure shorter lengths.

1 There are 100 cm in 1 m and 10 mm in 1 cm.

2 Take a look at this dog. It's 60 cm tall.

3 We can easily change this height into m, by dividing it by 100. So, the dog is 0.6 m tall.

4 We can even change this height into mm, by multiplying it by 10. This means the dog is 600 mm tall.

5 We usually use mm to measure much smaller things, like the bumblebee buzzing beside the dog. The bumblebee is 15 mm long.

Converting units of length

Length units are easy to convert. All we need to do is multiply or divide by 10, 100, or 1000.

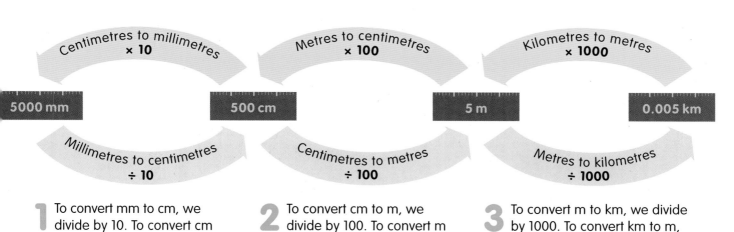

Centimetres to millimetres × 10

Metres to centimetres × 100

Kilometres to metres × 1000

5000 mm   500 cm   5 m   0.005 km

Millimetres to centimetres ÷ 10

Centimetres to metres ÷ 100

Metres to kilometres ÷ 1000

1 To convert mm to cm, we divide by 10. To convert cm to mm, we multiply by 10.

2 To convert cm to m, we divide by 100. To convert m to cm, we multiply by 100.

3 To convert m to km, we divide by 1000. To convert km to m, we multiply by 1000.

Calculating with length

Calculations with length measurements work just like
other calculations. You simply add, subtract, multiply, and divide
the numbers as you would usually.

Calculating with the same units

1 This tree is 16.6 m tall. Four years
ago, it was 15.4 m tall. How much
has it grown?

2 To find the difference in height,
we need to subtract the
smaller number from the larger
number: 16.6 − 15.4 = 1.2

3 This means that the tree has
grown 1.2 m in four years.

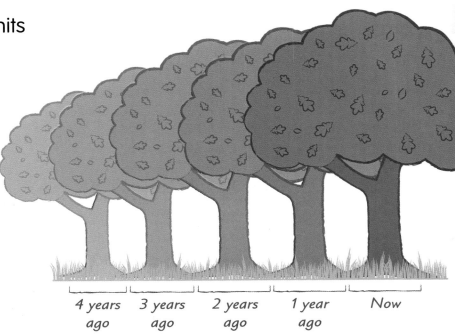

4 years   3 years   2 years   1 year   Now
ago   ago   ago   ago

4 Let's try a trickier problem.
We know the tree has
grown 1.2 m over four years,
but how much is that per year?

5 To solve this problem, all
we need to do is divide the
amount it has grown by the
number of years: 1.2 ÷ 4 = 0.3

6 So, the tree grew
0.3 m each year.

TRY IT OUT

Share the distance

This running track is
200 m long. If the four
robots each ran the same
distance in a relay race,
how far will each robot
need to run to cover the
whole track?

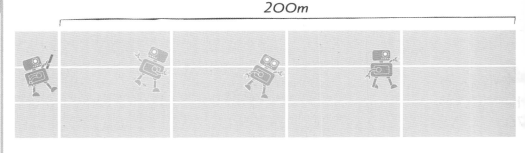

200m

1 To figure out the answer,
all you need to do is a
simple division calculation.

2 Just divide the length of
the track by the number of
robots sharing the distance.

Answer on page 319

Calculating with mixed units

We already know we can use different units to record length. If you are calculating with lengths, it is really important to make sure the values are all in the same unit before you start calculating.

> When calculating with distances, make sure the measurements are all in the same unit.

1 The robot in this picture is going to leave his house and travel 760 m to the toy shop, 1.2 km to the playground, and then 630 m to the zoo. How far is the total journey?

2 First, we have to put all the measurements into the same units. So, we need to change the distance between the toy shop and the playground from kilometres to metres.

3 Remember, to convert kilometres to metres, we just multiply the number of kilometres by 1000, because 1 km is the same as 1000 m:
1.2 × 1000 = 1200

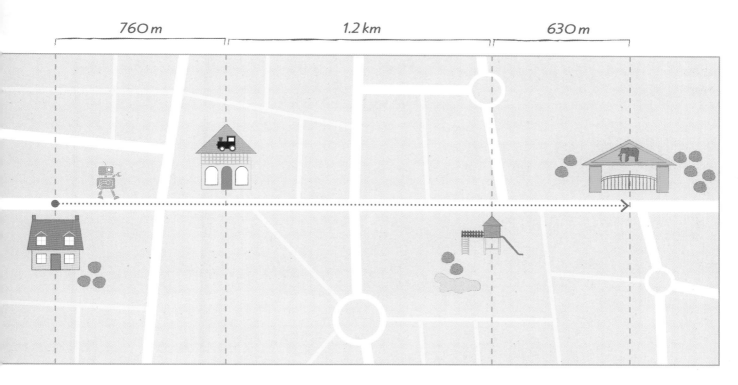

760 m   1.2 km   630 m

4 Now we can add all the distances together because they are all in metres:
760 + 1200 + 630 = 2590

5 2590 is quite a large number, so converting it back into kilometres will make it a more sensible number. To do this, we just need to divide by 1000: 2590 ÷ 1000 = 2.59

6 So, the robot will travel a total distance of 2.59 km.

Perimeter

Perimeter means the distance around the edge of a closed shape. If you imagine the shape is a field surrounded by a fence, the perimeter is the length of the fence.

The perimeter of a shape is the sum of the lengths of all its sides.

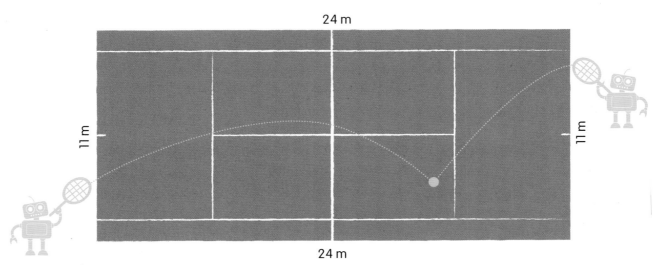

24 m

11 m

11 m

24 m

1 To find the perimeter of a shape, we need to measure the length of each side and add them all together.

2 We measure perimeter using the same units as we use to measure length. It is important that the sides are in the same unit when we add them all together.

3 Look at this tennis court. We can find the perimeter by adding up the length of each side:
11 + 24 + 11 + 24 = 70

4 This means that the perimeter of the tennis court is 70 m.

TRY IT OUT

Unusual shapes

We measure the perimeter of an unusual shape in the same way as a rectangle – just find the sum of all the sides. Can you add up the sides of these two shapes to find their perimeter?

Answers on page 319

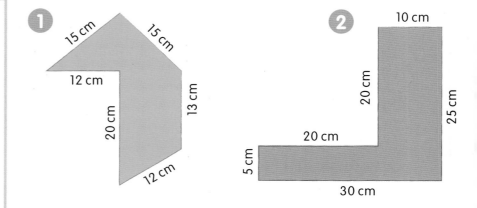

1
15 cm
15 cm
12 cm
20 cm
13 cm
12 cm

2
10 cm
20 cm
25 cm
20 cm
5 cm
30 cm

What if we don't know the lengths of all the sides?

Sometimes we don't know the lengths of all the sides of a shape. If a shape made up of one or more rectangles has a measurement missing, we can still figure out the missing length and the perimeter.

1 Look at this field. We need to find the perimeter but the length of one side is missing.

2 The field's corners are right angles, so its opposite sides are parallel to each other. That means if we know the length of one side, we can work out the length of the unknown part of the opposite side.

3 Let's find the missing length. The opposite side is 12 m long, so the two sides facing it must also have a total length of 12 m.

4 To find the missing length, we just need to subtract 9 from 12: $12 - 9 = 3$. So, the length of the missing side is 3 m.

5 Now we can work out the total perimeter by adding up the lengths of all the sides: $12 + 6 + 9 + 5 + 3 + 11 = 46$

6 This means that the perimeter of the field is 46 m.

6 m

9 m

12 m

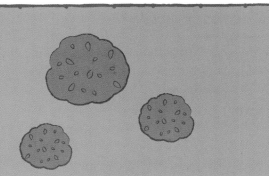

5 m

?

11 m

Using formulas to find perimeter

If we remember some basic facts about 2D shapes, we can use formulas to find their perimeters. These formulas use letters to represent the lengths of the sides. This makes it easier for us to remember how to calculate the perimeters of lots of different shapes.

Square

1 We know that all four sides of a square are the same length. We can find the perimeter by adding those four sides together.

2 Look at this red square. If we call the length of each side "a", we can say Perimeter = a + a + a + a. A simpler way of writing this is:

Perimeter of a square = 4a

3 Let's imagine that the square's four sides were each 2 cm long. The perimeter would be 8 cm, because 4 × 2 = 8

Rectangle

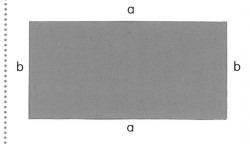

1 A rectangle has two pairs of opposite sides that are parallel and equal in length. Let's call the length in one pair "a" and the length in the other pair "b".

2 For a rectangle, we can add up the two lengths that are different then multiply by two, because there are two sides of each length. We use the formula:

Perimeter of a rectangle = 2 (a + b)

3 So, if the rectangle's sides were 2 cm and 4 cm long, the perimeter would be 12 cm, because 2 (4 + 2) = 12

Parallelogram

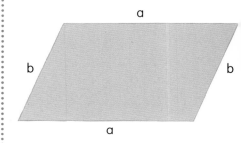

1 Just like a rectangle, a parallelogram has two pairs of opposite sides that are parallel and equal in length.

2 So, we can use the same formula for a parallelogram as for a rectangle, adding the two adjacent side lengths together then multiplying by two:

Perimeter of a parallelogram = 2 (a + b)

3 This means that if the sides were 3 cm and 5 cm, the perimeter would be 16 cm, because 2 (5 + 3) = 16

Using perimeter to find a missing measurement

If we know the perimeter of a shape and all of its side lengths except one, we can work out the length of the missing side with a simple subtraction calculation.

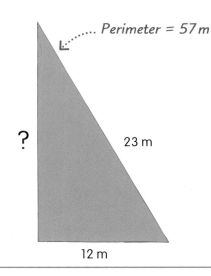

Perimeter = 57 m

? 　23 m

12 m

1 Look at this triangle. We know its perimeter and the lengths of two sides. Let's find the length of the unknown side.

2 We can find the length of the unknown side by simply subtracting the lengths that we know from the perimeter: 57 – 23 – 12 = 22

3 So, the unknown side is 22 m long.

Equilateral triangle

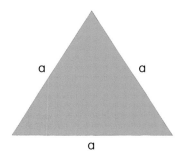

a　a
a

1 We know that an equilateral triangle has three sides that are all the same length.

2 Like we do with a square, we just need to multiply the length of one side by the number of sides. If we call the length "a", the formula we can use is:

Perimeter of an equilateral = 3a triangle

3 Let's imagine the three sides were each 4 cm long. The perimeter would be 12 cm, because 3 × 4 = 12

Isosceles triangle

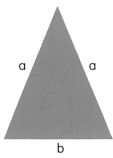

a　a
b

1 An isosceles triangle has two sides that are equal in length and one side that is different.

2 Let's call each of the two sides that are the same "a". To find the perimeter, we multiply "a" by two then add the length of the other side, "b":

Perimeter of an isosceles = 2a + b triangle

3 So, if the two sides that are equal in length were 4 cm and the different side was 3 cm, the perimeter would be 11 cm.

Scalene triangle

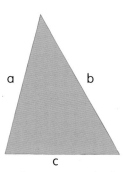

a　b
c

1 A scalene triangle has three sides that are all different lengths.

2 If we call the three sides "a", "b", and "c", we can find the perimeter by adding the three lengths together. We can use the formula:

Perimeter of a scalene = a + b + c triangle

3 So, if the triangle's sides were 4 cm, 5 cm, and 6 cm, then the perimeter would be 15 cm, because 4 + 5 + 6 = 15

Area

The amount of space enclosed by any 2D shape is called its area. We measure area using units called square units, which are based on the units we use for length.

We can find the area of a rectangle by dividing it into squares and counting the number of squares.

1 Look at this patch of grass. It is 1m long and 1m wide. We call it a square metre, and we write it like this: 1m².

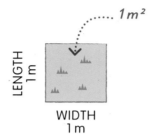

LENGTH 1m

WIDTH 1m

1m²

2 Take a look at this garden. We can work out its area by filling it with 1m² patches of grass and then counting the patches.

2 m

3 m

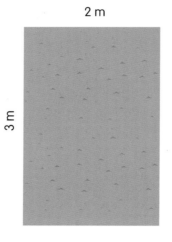

3 As the garden fills up, we can see that two squares will fit along its width and three along its length.

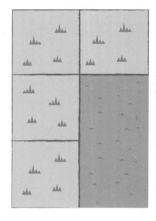

4 In total, we can fit exactly six 1m² patches into the garden. We can say it has an area of 6 m².

TRY IT OUT

Unusual areas

We can also use square units to work out the areas of more complicated shapes. Can you work out the areas of these shapes by counting the number of square centimetres in each one?

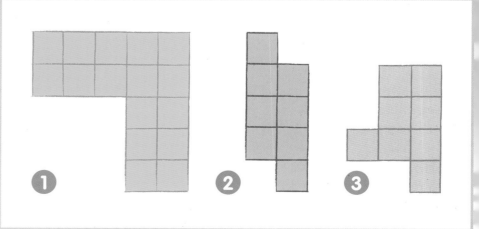

1   **2**   **3**

Answers on page 319

Estimating area

Finding the areas of shapes that are not squares or rectangles may seem tricky. But we can combine the number of completely full squares and partly full squares to estimate the area.

11 m

1 Look at this pond. Its unusual shape makes it difficult to work out its area.

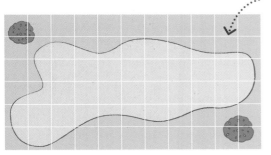

Each square is 1 m along each side

2 We can estimate its area if we draw a square grid over the pond, where each square represents 1 m².

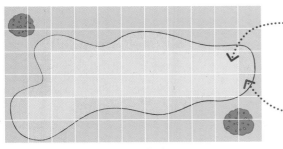

Count the squares that are completely filled with water

Ignore the squares that aren't completely filled

3 First, we count all the squares that are completely filled with water by colouring them in. There are 18 full squares.

Count the squares that are partially filled ....

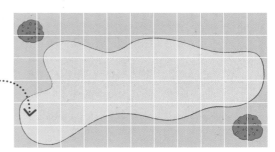

4 Next, we count the squares that are only partially filled by water. There are 26 partially filled squares.

5 Most of the partial squares cover just over or just under half a square. So, to estimate the number of squares they cover in total, we can divide the number by 2: 26 ÷ 2 = 13

6 Finally, we add together the areas of the full squares and the partially filled squares to get an estimate of the total area: 18 + 13 = 31

7 So, the area of the pond is approximately 31 m².

Drawing a square grid over an unusual shape can help us to find its estimated area.

Working out area with a formula

Using a formula is a much easier way to find a shape's area than having to count squares. Calculating with a formula means you can find the area of large shapes more quickly.

The area of a square or rectangle is: length × width

1 Look at this playground. We know that it has a width of 6 m and a length of 8 m.

2 If we put a square grid over the playground, we would see eight rows of six 1 m² units, making a total area of 48 m².

3 There is a quicker way to find the area than counting squares. We can use a formula.

4 If we multiply 6 by 8, we get 48. This is the same number as the number of metre squares we can fit into the playground.

5 We can write this as a formula that will work for any rectangle, including squares:

Area = length × width

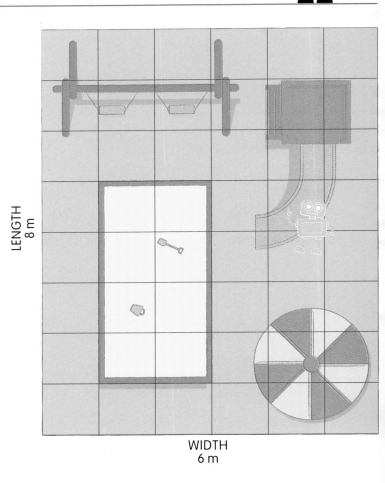

LENGTH
8 m

WIDTH
6 m

TRY IT OUT

See for yourself

The sandpit in the playground is 4 m long and 2 m wide. Can you use the formula to find the area of the sandpit?

Answer on page 319

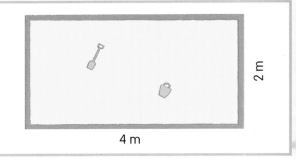

2 m

4 m

Area and missing measurements

Sometimes we know the length of one side of a rectangle and its area, but the length of the other side is unknown. To find the missing length, we simply need to use the numbers that we know in a division calculation.

1 To find a missing side length when we know the area, we just need to divide the area by the side length we do know.

2 This bedroom has an area of 30 m², and we know that it is 5 m wide. Let's figure out the length of the room.

3 To find the length, we divide the area by the width: 30 ÷ 5 = 6

4 This means that the room has a length of 6 m.

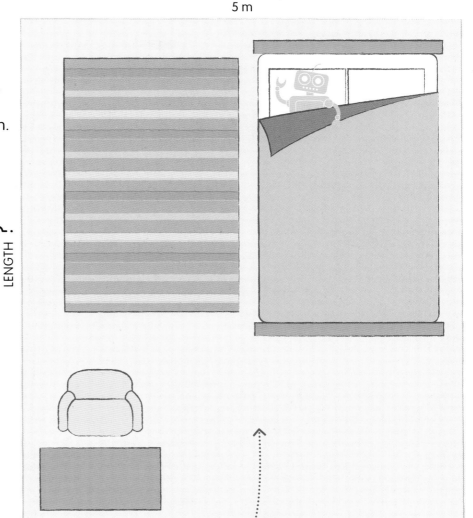

WIDTH
5 m

LENGTH ?

Area = 30 m²

Mystery length

Now you know how to find a missing length, see if you can do it yourself. This rug has an area of 6 m², and it is 2 m wide. How long is the rug?

Answer on page 319

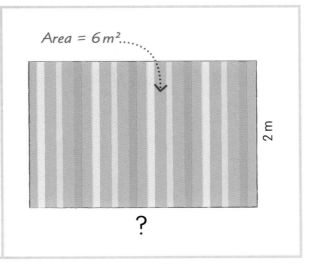

Area = 6 m²

2 m

?

When you know the area of a rectangle and the length of one side, you can find the length of the other side by dividing the area by the length you know.

Areas of triangles

Squares and rectangles aren't the only shapes with a
handy formula to help us work out their area. We can also use
formulas to find the areas of other shapes, including triangles.

The area of any triangle is:
½ base × height

Right-angled triangles

1 Look at this right-angled triangle. We're going
to use a formula to work out its area.

HEIGHT OF TRIANGLE

BASE OF TRIANGLE

.... *Make the triangle into a rectangle*

2 We can turn the triangle into a rectangle by
adding a second identical triangle. So, the
triangle takes up exactly half the rectangle's area.

3 We already know that the area of a rectangle
is: width × length. Here, the width of the
rectangle is equal to the base of the triangle, and
the length is equal to its height.

4 We also know the triangle has half the area
of the rectangle, so we can write a formula
for the area of a triangle like this:

Area of a triangle = ½ base x height

Other triangles

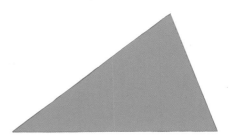

1 This scalene triangle looks a little trickier
to turn into a rectangle.

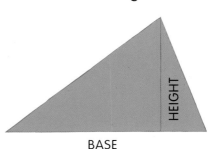

HEIGHT

BASE

2 First, draw a straight line down from the top
vertex to the base to make it into two right-
angled triangles.

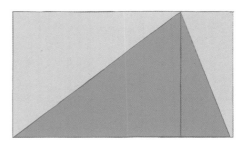

3 Now, it's easy to turn these two triangles into
rectangles like we did before. This triangle takes
up half the area, too. So, the formula is the same:

Area of a triangle = ½ base x height

Areas of parallelograms

Parallelograms aren't too different from rectangles – they're quadrilaterals with opposite sides that are parallel and equal in length. Because parallelograms are so like rectangles, we can use the same formula to work out their areas.

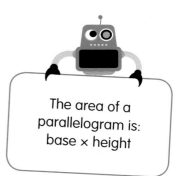

The area of a parallelogram is: base × height

1 Look at this parallelogram. Let's see why its area formula is the same as that of a rectangle.

2 First, let's draw a line straight down from the top corner of the parallelogram to its base. It creates a right-angled triangle.

Draw a straight line to make a triangle

HEIGHT

BASE

3 Imagine you could chop this triangle off and carry it over to the other end of the parallelogram.

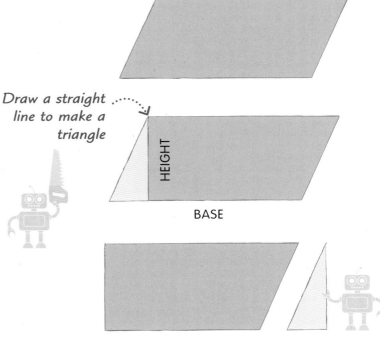

4 When you stick the triangle on the other end, it fits perfectly and makes the parallelogram into a rectangle.

5 This means that we can find the area by multiplying the height of the parallelogram by the length of the base, just like we did with the rectangle:

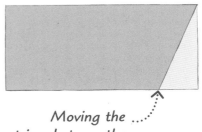

Moving the triangle turns the parallelogram into a rectangle

Area of a parallelogram = base x height

Areas of complex shapes

Sometimes you will be asked to find areas of shapes that look very complicated. Breaking these shapes into more familiar ones, like rectangles, makes finding the area much easier.

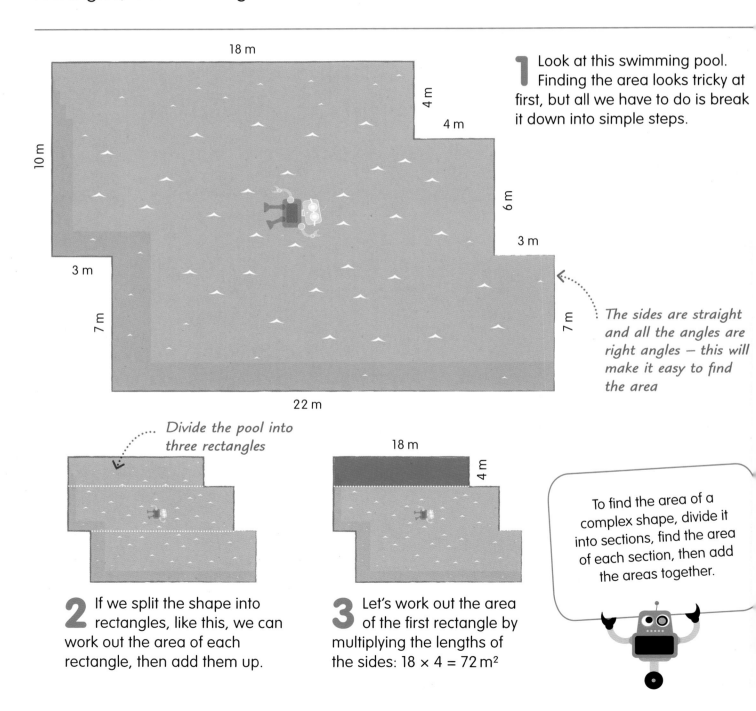

1 Look at this swimming pool. Finding the area looks tricky at first, but all we have to do is break it down into simple steps.

The sides are straight and all the angles are right angles – this will make it easy to find the area

18 m

4 m

4 m

10 m

6 m

3 m

3 m

7 m

7 m

22 m

Divide the pool into three rectangles

18 m

4 m

2 If we split the shape into rectangles, like this, we can work out the area of each rectangle, then add them up.

3 Let's work out the area of the first rectangle by multiplying the lengths of the sides: 18 × 4 = 72 m²

To find the area of a complex shape, divide it into sections, find the area of each section, then add the areas together.

Add two measurements together to find this length .....

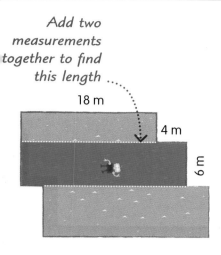

18 m

4 m

6 m

Now we know the area of all three parts of the pool .....

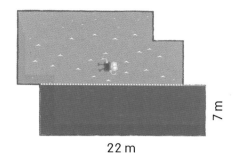

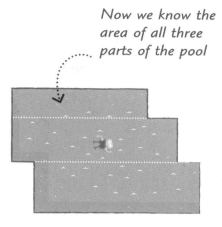

22 m

7 m

4 To find the area of the second rectangle, we first need to work out its length by adding 4 and 18 to get 22. Then we can multiply the lengths of the sides: $22 \times 6 = 132 \, m^2$

5 For the final section, we simply need to multiply the lengths of the sides to find its area: $22 \times 7 = 154 \, m^2$

6 All we need to do now is add together the three areas to get the pool's total area: $72 + 132 + 154 = 358$

7 So, the area of the swimming pool is $358 \, m^2$.

TRY IT OUT

How big is this room?

Now you know how to work out the area of a complex shape, can you find the total area of the floor of this room?

1 Begin by breaking up the floor into rectangles. There's more than one way to do this.

2 Once you've broken the shape up, you'll need to do some addition or subtraction to find some of the measurements you'll need.

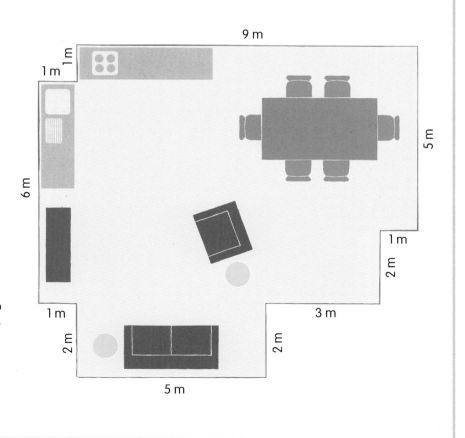

9 m

1 m

1 m

5 m

6 m

1 m

2 m

1 m

3 m

2 m

2 m

5 m

Answer on page 319

Comparing area and perimeter

We know how to find the area and perimeter of shapes, but how are they related? If two shapes have the same area, they don't always have the same perimeter. This is true the other way round, too.

Even if shapes have the same area, they may not have the same perimeter. Also, shapes with the same perimeter may not have the same area.

Same area but different perimeter

Look at these three zoo enclosures. They all have the same area – 240 m². Does this mean that they all have the same perimeter too?

1 If we look at the zebra enclosure, we can see that it has a perimeter of 62 m.

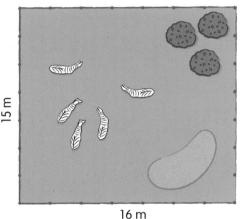

15 m

16 m

Perimeter = 62 m
Area = 240 m²

2 The perimeter of the penguin enclosure is 64 m. This is greater than the perimeter of the zebra enclosure, even though the area is the same.

12 m

20 m

Perimeter = 64 m
Area = 240 m²

3 The tortoise enclosure has an even greater perimeter. Its perimeter is 68 m.

4 It's important to remember that even if shapes have the same area, they may not have the same perimeter.

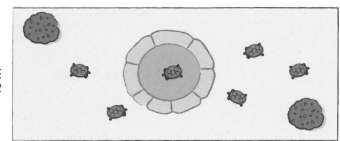

10 m

24 m

Perimeter = 68
Area = 240 m²

Same perimeter but different area

Now look at these two enclosures. They both have a perimeter of 80 m. Does this mean that they have the same area too?

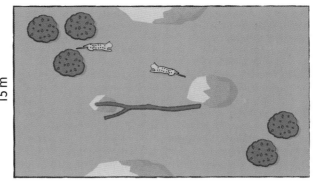

Perimeter = 80 m
Area = 375 m²

15 m

25 m

1 If we multiply the lengths of the sides of the leopard enclosure, we can see that its area is 375 m².

2 The area of the crocodile enclosure is 400 m². This is greater than the leopard enclosure, even though they both have the same perimeter.

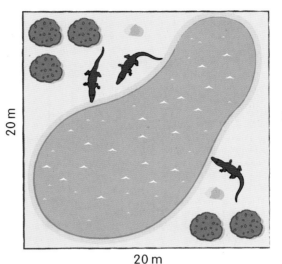

20 m

**Perimeter = 80 m
Area = 400 m²**

20 m

3 So, we can see that shapes with the same perimeter don't always have the same area.

Why aren't they the same?

When we change the measurements of a shape, why don't the perimeter and area change by the same amount? Perimeter is a measure of the length around the edge of a shape. Area is a measure of the space enclosed by the perimeter. This means that when we change one, the other isn't affected in the same way.

1 Take a look at this rectangle. If we keep the perimeter the same, but make it 1 cm longer and take 1 cm off the width, you might think the area would stay the same.

2 What's happened to the area and the perimeter? When we changed the shape, we removed 10 cm² from the bottom, but replaced it with only 3 cm² on the side.

3 So, the perimeter has stayed the same, but the area is now smaller.

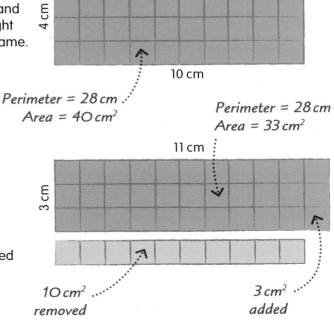

4 cm

10 cm

*Perimeter = 28 cm
Area = 40 cm²*

*Perimeter = 28 cm
Area = 33 cm²*

11 cm

3 cm

*10 cm²
removed*

*3 cm²
added*

Capacity

The amount of space inside a container is called its capacity. It is often used to describe how much liquid can be held in a container such as a water bottle. The capacity of a container is the maximum amount it can hold.

Volume is 50 l

CAPACITY (litres)

1 Capacity can be measured in units called millilitres (ml) and litres (l). There are 1000 ml in 1 l.

2 Millilitres are used to measure the size of small containers like a teacup (250 ml) or teaspoon (5 ml).

3 Litres are used to measure the sizes of larger containers like a large juice carton (1 l) or a bath (80 l).

4 Look at this fish tank. It has a capacity of 50 l.

Converting litres and millilitres

Converting between litres and millilitres is easy. To convert litres to millilitres, we multiply by 1000. To go from millilitres to litres, we divide by 1000.

1 To convert 5 l to millilitres, we multiply 5 by 1000. This gives the answer 5000 ml.

2 To convert the other way, from millilitres to litres, we divide 5000 ml by 1000, to give 5 l.

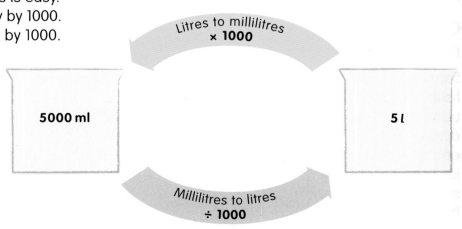

Litres to millilitres
× 1000

5000 ml

5 l

Millilitres to litres
÷ 1000

Volume

Volume is a measure of how big something is in three dimensions. Liquid volume is similar to capacity and is also measured in millilitres and litres. Adding and subtracting liquid volumes works just like other calculations.

1 Look at the fish tank again. We know that it has a capacity of 50 l, but it is now holding some water. The volume of the water is 10 l.

2 If a robot pours another 30 l of water into the tank, what will the volume of the water be now?

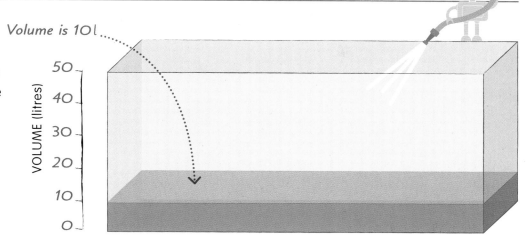

Volume is 10 l

3 To work out this sum, we simply have to add the two amounts together: 10 + 30 = 40

4 This means that the volume of the water in the tank is now 40 l.

Volume is 40 l

Calculating with mixed units

Sometimes you will have to do calculations using a mixture of different units. The easiest way to do this is to convert the units so that they are all the same.

1 This bottle of juice has a volume of 1.5 l. If you drink 300 ml of the juice, how much will be left in the bottle?

2 Changing the units of one of the amounts makes the calculation easier. Remember, to change litres to millilitres we multiply by 1000.

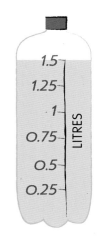

3 Let's change the bottle's volume to millilitres: 1.5 x 1000 = 1500

4 Now the calculation is simpler: 1500 − 300 = 1200

5 So, 1200 ml is left in the bottle.

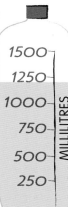

The volumes of solids

The volumes of 3D shapes are usually measured using units called cubic units. Cubic units are based on units of length, and they include cubic centimetres and cubic metres.

1 Look at this sugar cube. Each side of the cube is 1cm long, so we call it a cubic centimetre or 1cm³.

HEIGHT 1 cm

WIDTH 1 cm

LENGTH 1 cm

1 cm³

2 If each side was 1mm long, the volume would be 1mm³. If the sides were 1m long, it would be 1m³.

3 Now look at this box. We can work out its volume by filling it with cubic centimetres.

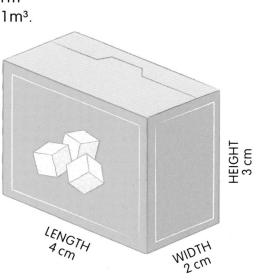

LENGTH 4 cm

WIDTH 2 cm

4 First, let's fill the bottom of the box. We can fit 8 cubic centimetres into this layer.

8 cm³

5 If we keep going until the box is full, we find the box holds 24 cubic centimetres. In other words, its volume is 24 cm³.

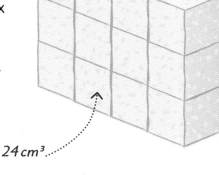

HEIGHT 3 cm

24 cm³

TRY IT OUT

Unusual shapes

You can use the method we've just learned to find the volume of all sorts of shapes, not just regular ones. Count the cubic centimetres to work out the volume of each of these three shapes.

Answers on page 319

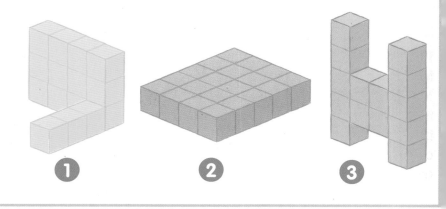

1   **2**   **3**

Working out volume with a formula

There is an easier way to work out the volumes of simple shapes like cuboids without having to count cubes. Instead, we can use a formula, calculating the number of units rather than counting them.

The volume of
a cube or cuboid is:
length × width × height

1 The volume of a cuboid can be written like this:

Volume of a cuboid = length x width x height

2 Let's work out the volume of this cereal packet.

3 First, we multiply the length by the width: 24 x 8 = 192

4 Next, we multiply the result by the height: 192 x 30 = 5760

5 This means that the volume of the packet is 5760 cm³.

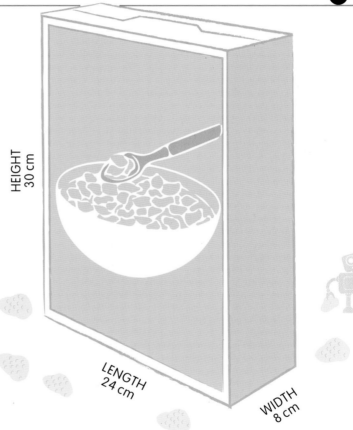

HEIGHT 30 cm

LENGTH 24 cm

WIDTH 8 cm

TRY IT OUT

Small things in big packages

This robot is going to cram a cardboard box full of 1 cm³ dice. The box has a volume of 1 m³. Can you work out how many dice will fit in the box using the formula? You might be surprised! Before you start your calculation, remember to convert the dimensions of the box into centimetres.

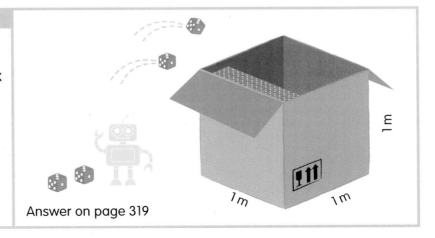

1 m

1 m

1 m

Answer on page 319

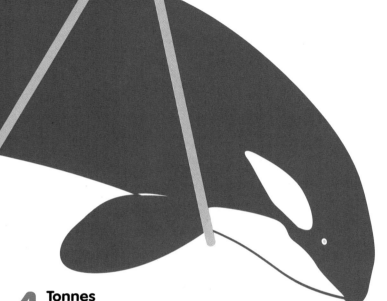

4 tonnes

Mass

Mass is the amount of matter, or material, contained within an object. We can measure mass using metric units called milligrams (mg), grams (g), kilograms (kg), and tonnes.

1 Milligrams
We measure very light things in milligrams. The mass of this ant is 7 mg.

7 mg

2 Grams
This frog has a mass of 5 g. There are 1000 mg in 1 g. 1 g is about the mass of a paperclip.

5 g

3 Kilograms
The mass of this big cat is 8 kg. There are 1000 g in 1 kg.

8 kg

4 Tonnes
Tonnes are used to measure very heavy things. This whale has a mass of 4 tonnes. 1 tonne is the same as 1000 kg.

Converting units of mass

Units of mass are easy to convert. We just have to multiply or divide by 1000 to switch between units.

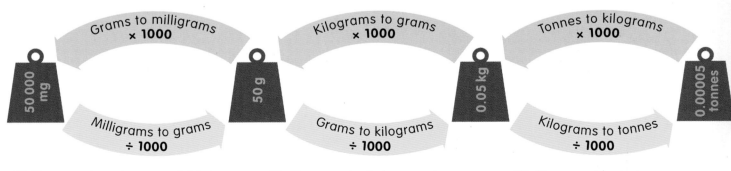

Grams to milligrams
× 1000

50 000 mg

Milligrams to grams
÷ 1000

50 g

Kilograms to grams
× 1000

Grams to kilograms
÷ 1000

0.05 kg

Tonnes to kilograms
× 1000

Kilograms to tonnes
÷ 1000

0.00005 tonnes

1 To convert mg to g, we divide by 1000. To convert back the other way, we multiply by 1000.

2 To convert g to kg, we also divide by 1000. To convert the other way, we multiply by 1000.

3 To convert kg to tonnes, we divide by 1000, too. To convert back to kg, we multiply by 1000.

Mass and weight

We often use the word weight when we mean mass but they're not actually the same. Weight is how hard the force of gravity attracts an object and is measured in a special unit called Newtons (N).

Mass is the amount of matter something is made up of. Weight is the amount of gravity acting on something.

1 If you were to travel around the Universe, your weight would change depending on where you were. This is because the gravity that acts on you is different in different places.

2 Even though your weight would be different, your mass would stay the same. This is because your mass is the amount of matter you are made up of, so it doesn't change.

ON EARTH
MASS 120 KG
WEIGHT 1200 N

3 Your mass and weight are almost the same everywhere on Earth. This astronaut's weight on Earth is 1200 N. Her mass is 120 kg.

ON THE MOON
MASS 120 KG
WEIGHT 200 N

ON JUPITER
MASS 120 KG
WEIGHT 2700 N

4 On the Moon, the astronaut weighs about one-sixth of what she weighs on Earth because the Moon's gravity is one-sixth of Earth's gravity.

IN SPACE
MASS 120 KG
WEIGHT 0 N

5 In outer space, there is no gravity, so even though our astronaut has no weight, she still has the same mass as she would have on Earth.

6 The astronaut would weigh more than twice as much on Jupiter compared to Earth because Jupiter's gravity is much stronger than Earth's. She would feel very heavy, but her mass would remain the same.

Calculating with mass

We can do calculations with mass in the same way that we do with lengths and other measurements. As long as the masses are in the same units, we can simply add, subtract, multiply, or divide them.

Calculating mass with the same units

85 g

1 Look at these three parrots. If we add their masses together, what is their total mass?

2 To work this out, we simply need to add the three masses together:
85 + 73 + 94 = 252

73 g

3 So, the parrots have a total mass of 252 g.

94 g

Comparing mass with mixed units

When you're tackling a problem that involves mass, it's important to pay attention to the units. If the masses are not all in the same unit, you'll need to start by doing some conversion. We looked at converting masses on page 182.

85 g

1 Look at these three animals. Can you put them in order, from the heaviest to the lightest?

2 It might seem tricky at first because their masses are not in the same unit. To make it easier, we're going to do a conversion.

3 Let's change the parrot's mass into kilograms so that all the masses are in the same unit – kilograms.

TRY IT OUT

Weighing it up

Subtracting with mass is just as easy as adding. Can you calculate how much heavier the yellow toucan is than the green toucan? All you need to do is subtract the smaller mass from the larger mass.

Answer on page 319

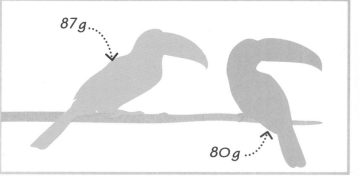

87 g

80 g

Remember, there are 1000 g in 1 kg and 1000 kg in 1 tonne.

35 kg

130 kg

4 To change 85 g to kilograms, we just divide by 1000:
85 ÷ 1000 = 0.085 kg

5 Now it's much clearer which order the animals go in, and we can order the numbers from largest to smallest.

6 The tiger has the largest mass of 130 kg, the snake has the next largest at 35 kg, and the parrot is the smallest at just 0.085 kg.

TRY IT OUT

Convert and calculate

Can you work out the total mass of this group of gibbons? Remember to take a careful look at the units.

1 First, you should convert the masses of the gibbons into the same unit.

2 Then, you simply add up their masses.

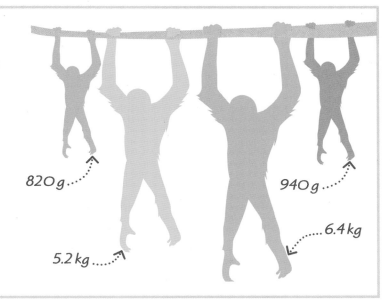

820 g

940 g

6.4 kg

5.2 kg

Answer on page 319

Temperature

Temperature is a measure of how hot or cold something is. We measure it using a thermometer and can record it in units called degrees Celsius (°C) or degrees Fahrenheit (°F). You might also hear degrees Celsius called degrees centigrade.

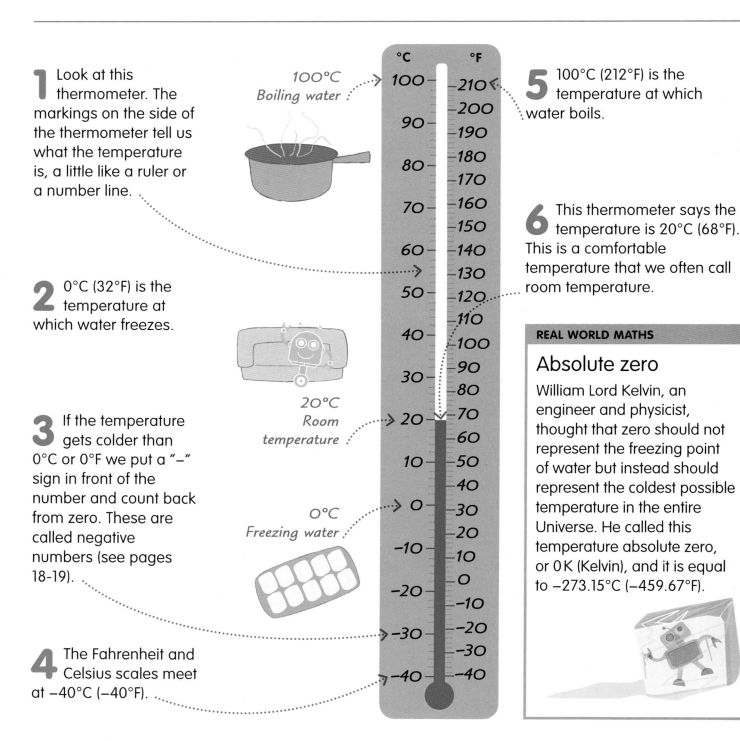

1 Look at this thermometer. The markings on the side of the thermometer tell us what the temperature is, a little like a ruler or a number line.

2 0°C (32°F) is the temperature at which water freezes.

3 If the temperature gets colder than 0°C or 0°F we put a "−" sign in front of the number and count back from zero. These are called negative numbers (see pages 18-19).

4 The Fahrenheit and Celsius scales meet at −40°C (−40°F).

100°C
Boiling water

20°C
Room temperature

0°C
Freezing water

5 100°C (212°F) is the temperature at which water boils.

6 This thermometer says the temperature is 20°C (68°F). This is a comfortable temperature that we often call room temperature.

REAL WORLD MATHS

Absolute zero

William Lord Kelvin, an engineer and physicist, thought that zero should not represent the freezing point of water but instead should represent the coldest possible temperature in the entire Universe. He called this temperature absolute zero, or 0 K (Kelvin), and it is equal to −273.15°C (−459.67°F).

Calculating with temperature

We can do addition and subtraction with temperatures measured in degrees Celsius and Fahrenheit, although we can't do multiplication or division.

> The scale on a thermometer works just like the scale on a number line.

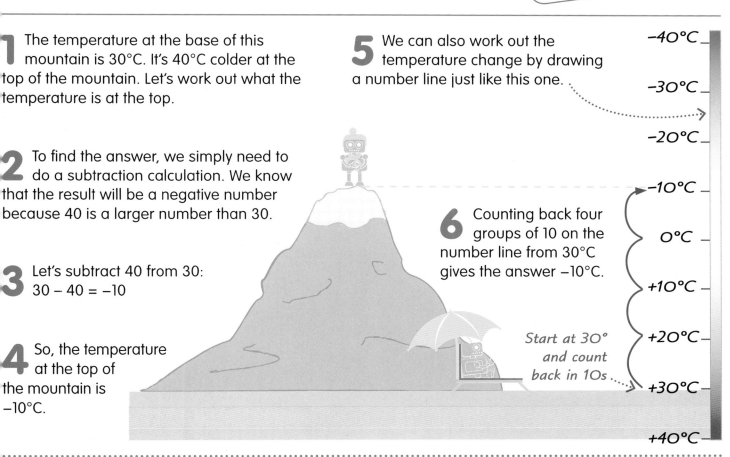

1 The temperature at the base of this mountain is 30°C. It's 40°C colder at the top of the mountain. Let's work out what the temperature is at the top.

2 To find the answer, we simply need to do a subtraction calculation. We know that the result will be a negative number because 40 is a larger number than 30.

3 Let's subtract 40 from 30:
30 – 40 = –10

4 So, the temperature at the top of the mountain is –10°C.

5 We can also work out the temperature change by drawing a number line just like this one.

6 Counting back four groups of 10 on the number line from 30°C gives the answer –10°C.

Start at 30° and count back in 10s

–40°C
–30°C
–20°C
–10°C
0°C
+10°C
+20°C
+30°C
+40°C

TRY IT OUT

World weather

In Sweden, the average temperature in February is –3°C. If it's 29°C hotter in India, what is the temperature?

Answer on page 319

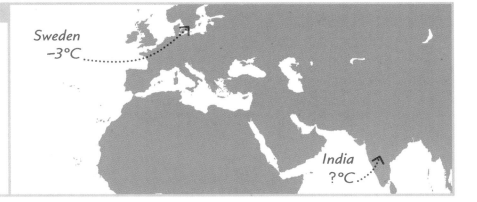

Sweden
–3°C

India
?°C

Imperial units

We've looked at the units we use to measure things in the metric system. In some countries, a different system is used to measure. It's called the imperial system, and it's useful to be aware of the different units that make up the system.

The imperial system

Each unit in the imperial system is very different to the next, because they have been inspired by different things over thousands of years.

1 Mass
Just as with the metric system, there is a range of different units we can use in the imperial system to measure mass, such as ounces, pounds, and imperial tons.

2 In the imperial system, we use units called pounds to weigh things like this dog.

3 The dog has a mass of 55 pounds.

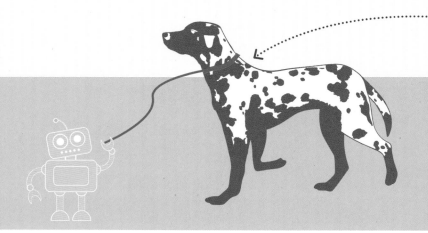

4 If we were weighing the dog in metric units, we would measure it in kilograms. The dog has a mass of about 25 kilograms.

REAL WORLD MATHS

Mars mix-up

In 1999, NASA made a very expensive mistake with units. Their $125 million Mars Climate Orbiter was lost because someone didn't do the right conversions! One team had been working in metric units, while the other worked in imperial units. As a result, the probe flew too close to Mars. It was lost, and probably destroyed, as it entered the planet's atmosphere.

Length

The imperial units for length and distance are called inches, feet, yards, and miles.

This tall building is 760 yards tall and is 1 mile away from the dog.

Measuring in metric units, we can say the building is about 690 metres tall and is 1.6 kilometres away from the dog.

Volume and capacity

There are two imperial units commonly used for volume and capacity: pints and gallons. This pond has a volume of 480 pints or 60 gallons. This is roughly the same as 270 litres.

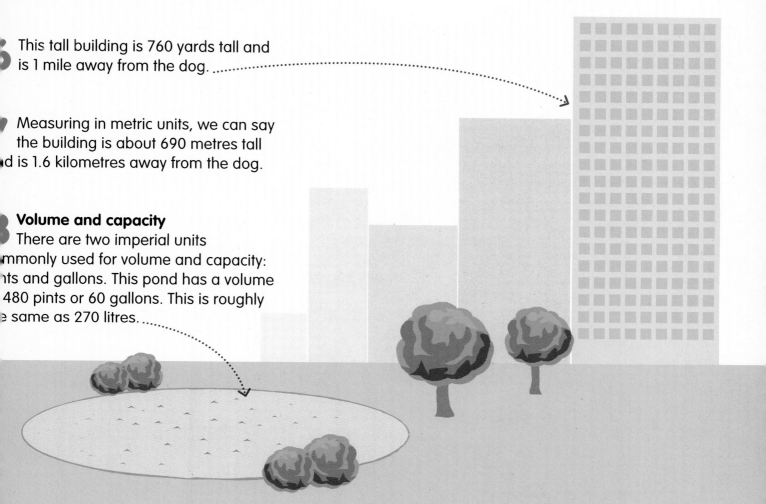

Converting between the imperial and metric systems

We have learned about converting measurements within the metric system, but we can also convert between imperial and metric units. It works both ways, and all we need is a number called a conversion factor.

1 Let's convert 26 metres into feet. All we need to do is multiply each 1 metre by its value in feet. We call this value the conversion factor.

2 1 m is equal to 3.3 ft, so the conversion factor we use to change metres to feet is 3.3.

3 Now we multiply 26 by the conversion factor: 26 × 3.3 = 85.8

4 So, 26 m is the same as 85.8 ft.

26 METRES

? FEET

$$26\,m = ?\,ft$$

$$26 \times 3.3 = 85.8$$

$$26\,m = 85.8\,ft$$

Imperial units of length, volume, and mass

Just like the metric system, the imperial system has many different units that we can use to measure length, volume and capacity, and mass. We looked at how this system compares with the metric system on pages 188-89.

Length

1 Length can be measured in imperial units called inches, feet, yards, and miles.

2 Look at this cat. We can measure its height in inches. The cat is 12 inches tall.

3 There are 12 inches in 1 foot, so we can also say that the cat is 1 foot tall.

4 Yards are used to measure longer distances. There are 3 feet in 1 yard, so the cat is 1/3 yard tall.

5 Miles are usually used to measure even longer distances, like the distance between two towns. There are 1760 yards in 1 mile.

Volume and capacity

1 Volume and capacity can be measured in imperial units called pints and gallons. We can also use cubic imperial units, such as cubic inches and cubic feet. We looked at cubic units on pages 180-181.

2 Look at this fish tank. We can measure its capacity in pints. The capacity is 88 pints.

3 We can also measure capacity in an imperial unit called gallons. There are 8 pints in 1 gallon, so we usually use this unit to measure larger containers or volumes of liquid.

4 We could say the fish tank has a capacity of 11 gallons.

Mass

1 We can measure the mass of very light things in an imperial unit called ounces. This bird has a mass of 3 ounces.

2 We can also use pounds to measure mass. This big cat has a mass of 18 pounds. There are 16 ounces in 1 pound.

3 The imperial ton is used to measure very heavy things. There are 2240 pounds in an imperial ton. This elephant has a mass of 3 imperial tons. A very similar unit is used in the metric system, called tonnes or metric tons. It has a slightly different mass to the imperial ton.

Imperial and metric units

Imperial and metric units aren't very closely related. This table shows the equivalent measurements in metric and imperial to give you an idea of how the units relate to each other.

LENGTH	
1 inch = 2.54 centimetres	1 centimetre = 0.39 inch
1 foot = 0.30 metres	1 metre = 3.28 feet
1 yard = 0.91 metre	1 metre = 1.09 yards
1 mile = 1.61 kilometres	1 kilometre = 0.62 mile

VOLUME AND CAPACITY	
1 pint = 0.57 litre	1 litre = 1.76 pints
1 gallon = 4.55 litres	1 litre = 0.22 gallons

MASS	
1 ounce = 28.35 grams	1 gram = 0.04 ounce
1 pound = 0.45 kilogram	1 kilogram = 2.20 pounds
1 imperial ton = 1.02 tonnes	1 tonne = 0.98 imperial ton

Telling the time

We measure the passage of time to organise our everyday lives. Sometimes we need to know how long something takes, or we need to be in a certain place at a particular time. We use seconds, minutes, hours, days, weeks, months, and years to measure time.

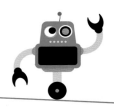

If we're writing the time using the 12-hour clock, we write a.m. or p.m. to show whether it's morning or afternoon.

Clocks

1 Look at this clock. The numbers around the edge help us measure which hour of the day it is. There are 24 hours in a day – 12 in the morning and 12 in the evening.

2 The shortest hand on the clock is the hour hand. It points to which hour of the day it is.

3 The marks around the edge of the clock tell us the minutes of an hour. There are 60 minutes in one hour.

The hands rotate in this direction, cal clockwise

4 There are no numbers to tell us precisely which minute it is. Instead, we use the hour numbers to help us count up in fives to work it out. The longer hand points to the minutes.

5 There are 60 seconds in a minute. Some clocks have a long, thin second hand that moves quickly around the clock face – one full turn takes one minute.

Types of clocks

Not all clocks look like the one above. Some clocks don't have hands at all. Others show all 24 hours in the day, instead of just using the numbers 1 to 12.

Sometimes the number 4 is written "IIII"

1 Some clocks use Roman numerals to mark the hours. We looked at Roman numerals on pages 10-11.

2 24-hour clocks have extra numbers to count up from 12 to 24, because there are 24 hours in a day.

Hours   *Minutes*

15 : 27

3 Digital clocks have no hands. They tell us the time with digits. They often use the 24-hour clock.

Reading the time

We describe the time by saying which hour of the day it is and how many minutes of that hour have passed. We can describe the number of minutes past the hour that's just gone, or how many minutes it is to the next hour.

5 minutes have gone by since 4 o'clock

The minute hand is halfway round the clock, so it is half past the hour

1 On the hour
When the minute hand is pointing to 12, the time is on the hour. We use the word "o'clock". This clock is showing 8 o'clock.

2 Half past an hour
When the minute hand points to 6, the time is halfway through an hour. The time on this clock is half past two.

3 Minutes past an hour
We ususally describe other times in multiples of 5, instead of being very precise. The time on this clock is 5 past 4. That means it's 5 minutes after 4 o'clock.

There are 15 minutes left until the next hour

4 Quarter past an hour
We can split hours into quarters. When the minute hand points to 3, we say it's quarter past the hour. This clock is showing quarter past ten.

5 Quarter to an hour
Here the minute hand is pointing to 9. Instead of saying it is three-quarters past, we say it's quarter to the next hour. The time on this clock is quarter to seven.

6 Minutes to an hour
When the minute hand goes past the number 6, we say how many minutes it is until the next hour. This clock is showing 10 to 5.

Converting seconds, minutes, hours, and days

There are 60 seconds in a minute, 60 minutes in an hour, and 24 hours in a day. So, converting time is harder than for other units where we can multiply or divide by 10, 100, or 1000.

Minutes to seconds **× 60**
Hours to minutes **× 60**
Days to hours **× 24**

| 21600 sec | 360 min | 6 h | 0.25 days |

Seconds to minutes **÷ 60**
Minutes to hours **÷ 60**
Hours to days **÷ 24**

1
To convert 21600 seconds to minutes, we divide by 60, which gives 360 minutes. To convert minutes to seconds, we multiply by 60.

2
To convert 360 minutes to hours, we divide by 60, which gives 6 hours. To convert hours to minutes, we multiply by 60.

3
To convert 6 hours to days, we divide by 24, which gives 0.25 days. To convert days to hours, we multiply by 24.

Dates

As well as seconds, minutes, and hours, we can measure time in units called days, weeks, months, and years. We use these units to measure periods of time that are longer than 24 hours.

One year is 365 days long, except on a leap year when there are 366 days.

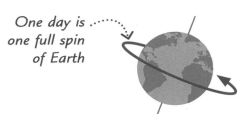

One day is one full spin of Earth

One week is one-quarter of the time between one full Moon and the next

1 Days
There are 24 hours in a day. A day is the length of time it takes for Earth to spin once on its axis.

2 Weeks
Days are grouped into a unit of time called weeks. There are 7 days in a week. This might be because it's a quarter of the cycle of the Moon (the time between one full Moon and the next).

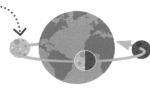

One month is based on the cycle of the Moon

One year is how long it takes for Earth to orbit the Sun

3 Months
There are between 28 and 31 days in a month. Months may have come from the lunar calendar at first, but have changed over time. Not all months have the same number of days.

4 Years
There are 365 days in a year. This is the same as 52 weeks or 12 months. A year is the length of time it takes for Earth to orbit the Sun once.

How long is a month?

To help us calculate with time, it's useful to know how many days there are in each month. Most of the months of the year have 30 or 31 days. February usually has 28 days, except in a leap year, when there are 29.

1 Look at these knuckles. The first 7 knuckles and the dips between them are labelled with a month.

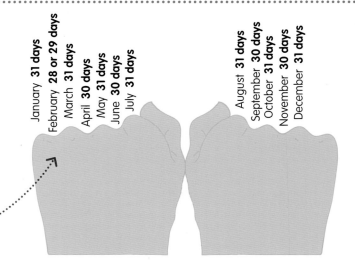

January **31 days**
February **28 or 29 days**
March **31 days**
April **30 days**
May **31 days**
June **30 days**
July **31 days**
August **31 days**
September **30 days**
October **31 days**
November **30 days**
December **31 days**

2 The months that sit on a knuckle are 31 days long: January, March, May, July, August, October, and December.

3 All the months, except February, that sit in a dip between two knuckles are 30 days long: April, June, September, and November.

Calendars

We use calendars to arrange all the days in a year into months and weeks. They help us to measure and keep track of the passing of time.

This January begins on a Friday and ends on a Sunday.

January

M	T	W	T	F	S	S
				1	2	3
4	5	6	7	8	9	10
11	12	13	14	15	16	17
18	19	20	21	22	23	24
25	26	27	28	29	30	31

February will begin on a Monday.

February

M	T	W	T	F	S	S
1	2	3	4	5	6	7
8	9	10	11	12	13	14

1 Look at this calendar showing the month of January.

2 The 365 days in a year don't fit neatly into a perfect number of weeks or months, so the day of the week that a month begins and ends on changes each year.

3 Here, January starts on a Friday and ends on a Sunday. This means that the previous month, December, ended on a Thursday and the next month, February, will begin on a Monday.

4 In following years, January will begin and end on different days of the week.

5 When we want to refer to a specific day in the year, or date, we say the number of the day in the calendar, followed by the month and year.

6 So, we can refer to the last day in January on this calendar as Sunday 31 January.

Converting days, weeks, months, and years

There are 7 days in a week and 12 months in 1 year, so converting these units of time can be quite tricky. Converting days or weeks to months is much harder, because the number of days and weeks in each month varies.

Weeks to days × 7

 42 days

 6 weeks

 48 months

 4 years

Days to weeks ÷ 7

Years to months × 12

Months to years ÷ 12

1 To convert 42 days to weeks, we divide by 7, to give 6 weeks. To convert back again, from weeks to days, we just multiply by 7. This takes us back to 42.

2 To convert 48 months to years, we divide by 12, which gives 4 years. To convert back the other way, we just multiply by 12. This takes us back to 48 months.

Calculating with time

It's simple to add, subtract, multiply, or divide an amount of time. As with other measurements, we just need to make sure the numbers are in the same units.

> When calculating with time, make sure you convert the times so that they are all in the same unit before you calculate.

Calculating time with the same units

If times are measured in the same units, it's easy to add and subtract them. But when we count on from a start time, we have to remember to count up to the nearest minute, hour, or day and then add on any remaining time.

1 It's 2.50 p.m. A robot is going to go on the big wheel, then walk to the exit of the fairground. Let's calculate what time it will be when the robot gets to the exit.

2 First, we need to add up the time for each part of the journey. The queue for the big wheel is 8 minutes long, the ride lasts 6 minutes, and it takes 2 minutes to walk to the fairground exit. Let's add these times up: 8 + 6 + 2 = 16

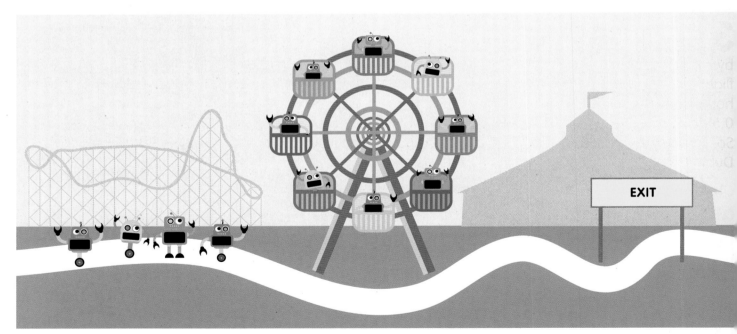

EXIT

3 Next, we add on minutes to 2.50 p.m. to take us to the next hour. Adding 10 minutes to 2.50 p.m. takes it to 3 p.m.

4 Finally, we add on the 6 minutes that are left over, taking the time to 6 minutes past 3.

5 So, the robot reaches the exit of the fairground at 3.06 p.m.

Comparing time with mixed units

Sometimes we're asked to calculate times that are in a mixture of units. We need to be careful to make sure the numbers are in the same unit before we start calculating.

1 Look at the times of these three flights from New York. Let's compare the duration of each journey and work out which is the shortest flight.

2 It's difficult to see which is shortest when the time for each journey is in a different unit. Let's convert them all into hours to make it simpler to work out.

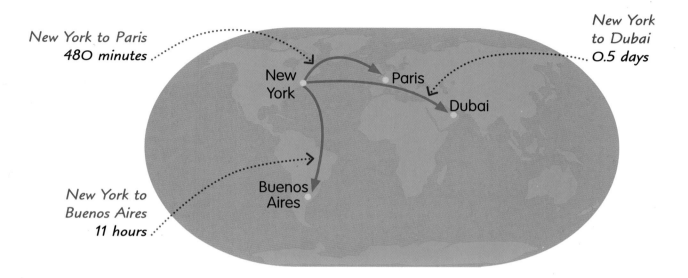

New York to Paris
480 minutes

New York
to Dubai
0.5 days

New
York

Paris

Dubai

New York to
Buenos Aires
11 hours

Buenos
Aires

3 The flight to Buenos Aires is already in hours, so we start by converting the duration of the flight to Dubai. There are 24 hours in a day, so we multiply 0.5 days by 24: 0.5 × 24 = 12. So, the journey from New York to Dubai takes 12 hours.

4 Next, we convert the time taken for the Paris flight into hours. We work this out by dividing by 60, because there are 60 minutes in an hour: 480 ÷ 60 = 8. So, it takes 8 hours to fly from New York to Paris.

5 We have worked out that it takes 8 hours to reach Paris, 11 hours to reach Buenos Aires, and 12 hours to reach Dubai from New York. So, the journey to Paris is shortest.

TRY IT OUT

Working with time

These robots are watching a film that is two and a half hours long. They have watched 80 minutes. How many minutes of the film are left?

1 First, convert the length of the film into minutes.

2 Now all you need to do is subtract the number of minutes watched from the total length of the film.

THE END

Answer on page 319

Money

Understanding money helps us to work out how expensive things are and check our change when we go shopping. Lots of systems of money (called currencies) are used around the world. In the UK, we use currency called pounds and pence.

1 Let's look at the items in this shop and see how the prices have been written.

2 We write a "£" sign in front of an amount in pounds or a "p" after amounts in pence.

3 £1 is equal to 100p. We call pounds a decimal currency, and we can think of amounts as decimal fractions.

4 We don't write £ and p together. If an amount is more than 99p, we just write the amount in pounds. The pence can be written as a decimal fraction of a pound.

5 So, one pound and forty-six pence is written £1.46.

6 An amount less than one pound is written with the letter "p" for pence. So, fifty-nine pence is written 59p.

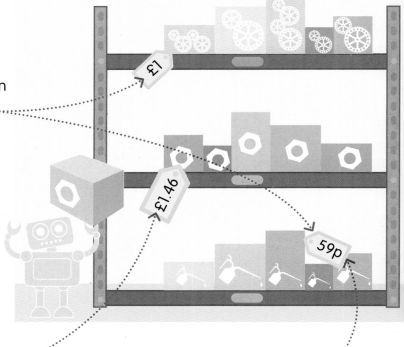

Converting units of money

Converting between pounds and pence is simple, because there are 100p in £1. To convert pence to pounds, we divide by 100. To convert pounds to pence, we multiply by 100.

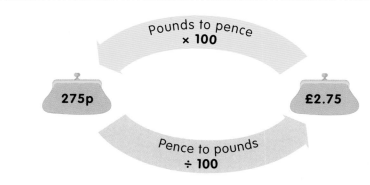

Pounds to pence × 100

275p   £2.75

Pence to pounds ÷ 100

1 To convert 275p to pounds, we divide 275 by 100. This gives the answer £2.75.

2 To convert the other way, from pounds to pence, we multiply 2.75 by 100 to give 275p.

Using money

In the UK, our money is made up of eight different coins (1p, 2p, 5p, 10p, 20p, 50p, £1, and £2) and four different notes (£5, £10, £20, and £50). We can mix them and swap them to make any amount of money we like.

---

1 Here are all the coins we can use to make different amounts. Let's see how we can combine these coins in different ways to make a total of £1.27.

2 We use the least number of coins if we combine the largest coin amounts possible: £1, 20p, 5p, and 2p.

3 We could combine other amounts to get the same total: 50p, 50p, 10p, 10p, 5p, 1p, and 1p.

4 We could even make £1.27 out of 127 1p coins! There are many different combinations we can use.

5 If we were in a shop, we could also pay with more than £1.27 and receive change. For example, we could pay with a £2 coin and receive 73p change.

REAL WORLD MATHS

Ancient money

Throughout history, people have used all sorts of things as money, like cowrie shells, elephant tail hairs, feathers, and whale teeth, because they were considered to be valuable.

We can combine notes with coins to make different amounts.

Calculating with money

We calculate with money in just the same way as we calculate with decimal numbers. We can learn to do this in our heads, using what we know about numbers, or use written methods, like column addition (see pages 86-87) and column subtraction (see pages 96-97).

Adding amounts of money

1 Let's add £26.49 and £34.63 using column addition. We looked at how to do column addition on pages 86-87.

$$£26.49 + £34.63 = ?$$

2 First, we write one number above the other number. Line up the decimal points, and put another decimal point lined up below in the answer line.

	£			p	
	^{1}2	^{1}6 .	^{1}4	9
+	3	4 .	6	3
	6	1 .	1	2

Line up the decimal points

3 Next, we work from right to left, adding each of the digits. The answer is £61.12

4 So, £26.49 + £34.63 = £61.12

$$£26.49 + £34.63 = £61.12$$

Round it up

Another way we can calculate with money is by rounding up or down. Prices are often close to a whole number of pounds, so it's simpler to round the amount up to work out the rough total. Then we just have to adjust the answer at the end. Remember, £1 is equal to 100p.

1 Let's add £39.98 and £45.99 by rounding both numbers up to the nearest whole pound.

$$£39.98 + £45.99 = ?$$

2 First, we add 2p to £39.98 to get £40 and add 1p to £45.99 to get £46. So, we've added a total of 3p.

$$£40 + £46 = ?$$

3 Next, we add the two amounts together: £40 + £46 = £86

$$£40 + £46 = £86$$

4 Finally, we just have to subtract the 3p that we added on at the start: £86 − 3p = £85.97

$$£86 - 3p = £85.97$$

5 So, £39.98 + £45.99 = £85.97

$$£39.98 + £45.99 = £85.97$$

Giving change

When we're paying for things, it's useful to be able to work out how much change we're owed. All we need to do is find the difference between the price of the items and the amount we paid. We do this by counting up. If the amounts aren't all in the same unit, we'll need to start by doing a conversion.

1 Look at these animals. Let's work out how much change we would get if we paid for three hamsters and one rabbit with a £10 note.

2 First, we need to find the total cost of the animals in pounds. We know 80p is the same as £0.80, so: (0.80 × 3) + 2.70 = 2.40 + 2.70 = 5.10. The animals cost £5.10 in total.

3 Now we can work out the change from £10. First, add on pence to take us to the nearest pound. Adding 90p to £5.10 takes us to £6.

4 Next, we add on pounds to take us up to £10. Adding £4 takes the total to £10.

5 Now, we add these two amounts together: £4 + 90p = £4.90

6 So, the change we get from buying the animals with a £10 note is £4.90.

Hamsters 80p each

Rabbits £2.70 each

TRY IT OUT

Calculate the cost

Can you work out the total cost in pounds of all these items? Remember to convert the amounts so they are all in the same unit.

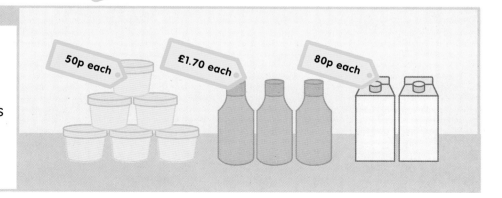

50p each

£1.70 each

80p each

Answer on page 319

In geometry, we study lines, angles, shapes, symmetry, and space. We can see plenty of geometric patterns in nature, such as the shapes of crystals and the symmetry of snowflakes. Geometry also has many other uses in everyday life – for example, we use it to navigate on journeys and to design and build structures such as bridges and buildings.

GEOMETRY

What is a line?

A line joins two points together. In geometry, lines can be either straight or curved. A line has a length that you can measure, but it has no thickness.

We call lines one-dimensional. They have length but no thickness.

1 Look at the straight line between A and B. It shows us the shortest distance between the two points.

A   **B**

2 The curved line bends round the trees, making the line between A and B longer than the straight line.

A   **B**

Prove it!

This map shows three possible routes between points A and B. Here is an easy way to prove that the straight line between the two is the shortest route.

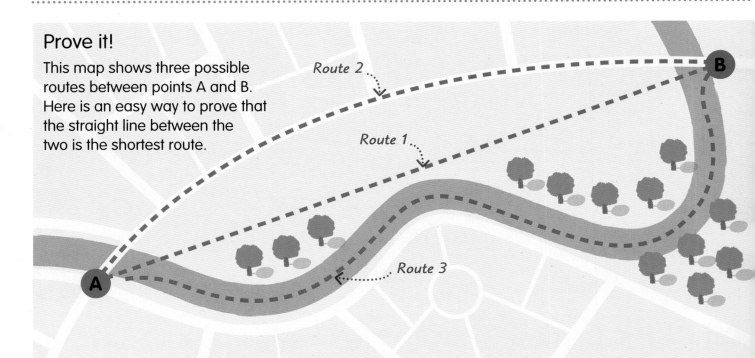

Route 2

Route 1

Route 3

B

A

1 Route 1 is a straight path. Stretch a piece of string from Point A to Point B along the path. Make a mark on the string where it reaches Point B.

2 Now do the same for Route 2, and mark where the string reaches Point B. The new mark is further along the string, so Route 2 must be longer than Route 1.

3 Now put the string along Route 3, the river. The mark you make this time will be the furthest along the string. So, Route 3 is the longest route.

Horizontal and vertical lines

We give lines different names to describe things about them, such as their direction or how they relate to other lines. Horizontal lines are level and go from side to side, while vertical lines go straight up and down.

1 A horizontal line goes from side to side, like the wings of this plane. It is parallel (level) with the horizon.

2 The struts that join the wings are vertical. They go up and down, and they're at right angles to the horizon.

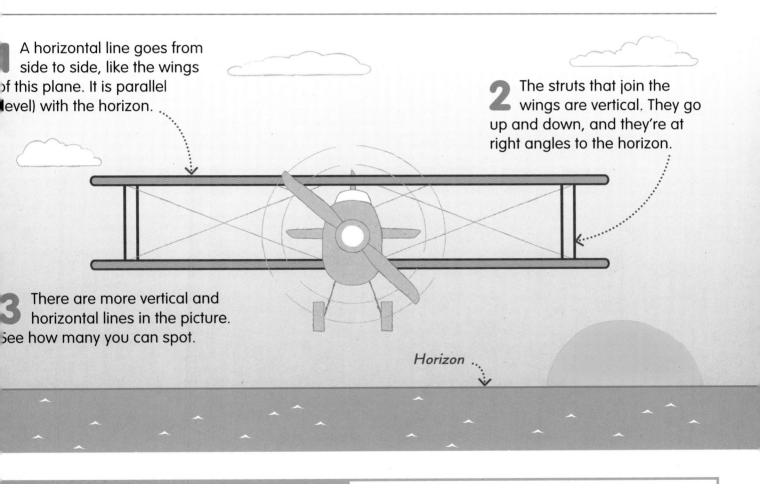

3 There are more vertical and horizontal lines in the picture. See how many you can spot.

Horizon

REAL WORLD MATHS

Horizontal or not?

A horizontal line is completely level. Some things need to be horizontal, such as bookshelves or the layers of bricks in the wall of a house. If a road has even a very gentle slope, a car would roll down to the bottom, unless we remembered to put the handbrake on!

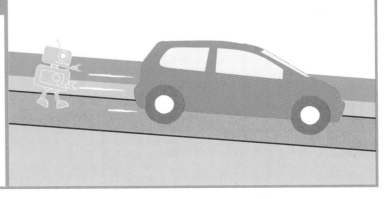

Diagonal lines

A straight line that slants is called a diagonal line. A diagonal line is not vertical or horizontal. Another name for a diagonal line is an oblique line.

> Straight lines can be horizontal, vertical, or diagonal.

1 Look at this picture of a zip wire. It is made of horizontal, vertical, and diagonal lines.

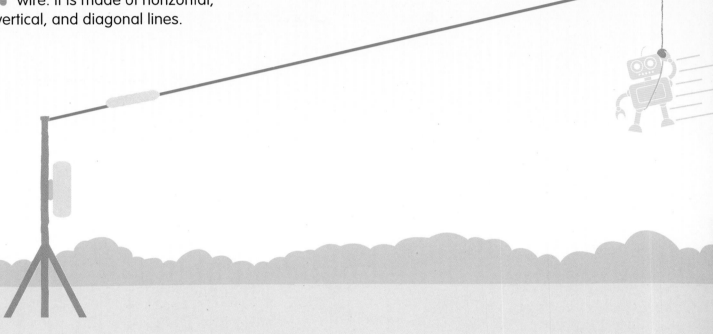

Diagonals inside shapes

In geometry, the word diagonal has another, more exact meaning. A diagonal is a straight line inside a shape. It joins two corners that are not next to each other.

1 Here are some examples of a diagonal inside a shape. We have shown one diagonal on each shape.

2 The more sides a shape has, the more diagonals it will have.

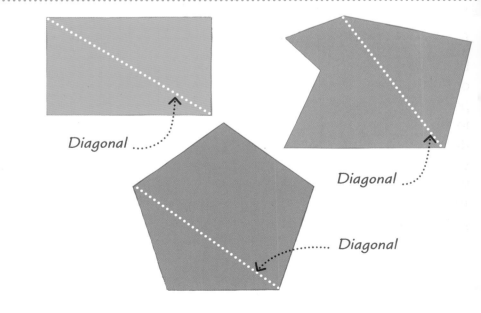

Diagonal

Diagonal

Diagonal

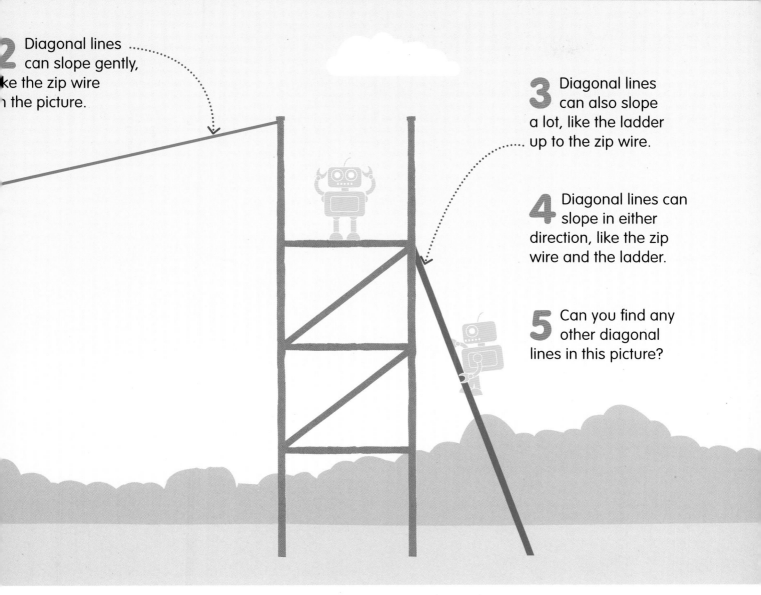

2 Diagonal lines can slope gently, like the zip wire in the picture.

3 Diagonal lines can also slope a lot, like the ladder up to the zip wire.

4 Diagonal lines can slope in either direction, like the zip wire and the ladder.

5 Can you find any other diagonal lines in this picture?

TRY IT OUT

Make a pattern with diagonals

Draw a regular hexagon (six-sided shape) or trace this one. Then use a ruler and pencil to draw diagonals from each corner to the other corners. This picture has three diagonals drawn for you, in white. When you have drawn all the lines, how many diagonals can you count inside the shape? Turn to page 320 to check your finished picture, then colour it in to make a pattern.

Answer on page 320

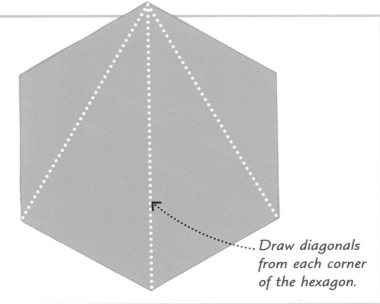

Draw diagonals from each corner of the hexagon.

Parallel lines

When two or more lines are exactly the same distance from each other all along their lengths, they are called parallel lines.

You can't have just one parallel line. They always come in sets of two or more.

1 Parallel lines
These ski tracks are parallel. No matter how long you make the lines, they will never meet, or intersect.

Parallel lines would never meet, even if the lines continued forever

2 Non-parallel lines
These tracks are not parallel. You can see that they are not the same distance from each other all along their lengths. If the tracks carried on, they would meet at one end.

At this end, the non-parallel lines get further apart the longer they continue

3 Curved parallel lines
Parallel lines can be wavy like these tracks, or zig-zag. What matters is that they are always the same distance apart, or equidistant, and never meet.

4 When lines are parallel, we mark them with small arrowheads, like this:

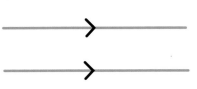

TRY IT OUT

Are they parallel?

Look at this scene. It's made up of several sets of parallel and non-parallel lines. Can you spot them all?

Answer on page 320

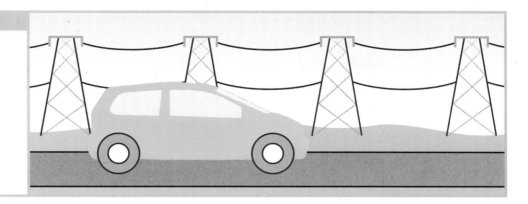

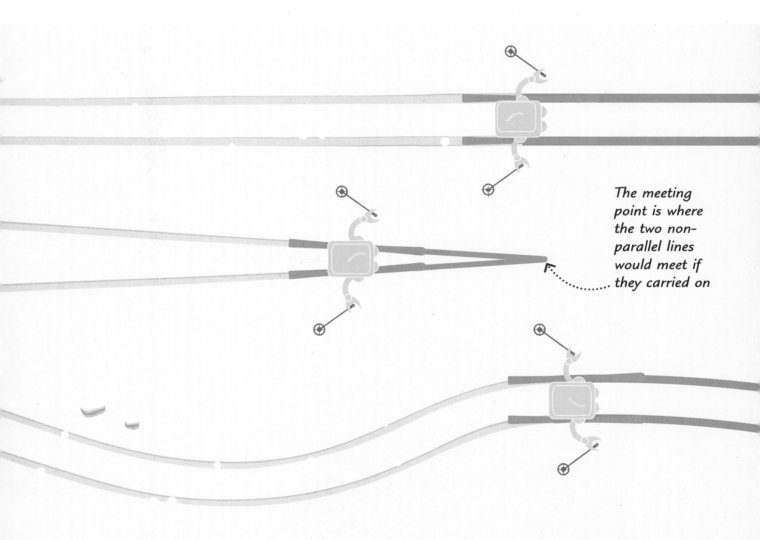

The meeting point is where the two non-parallel lines would meet if they carried on

5 Parallel lines don't just come in pairs – more than two lines can be parallel to each other. Parallel lines don't have to be the same length either.

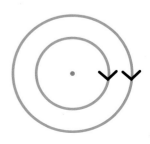

6 Lines that join up to make circles can also be parallel, like these circles with the same centre, called "concentric" circles.

Perpendicular lines

Perpendicular lines come in pairs. We call lines perpendicular when they are at right angles to each other. You can find out all about right angles on page 232.

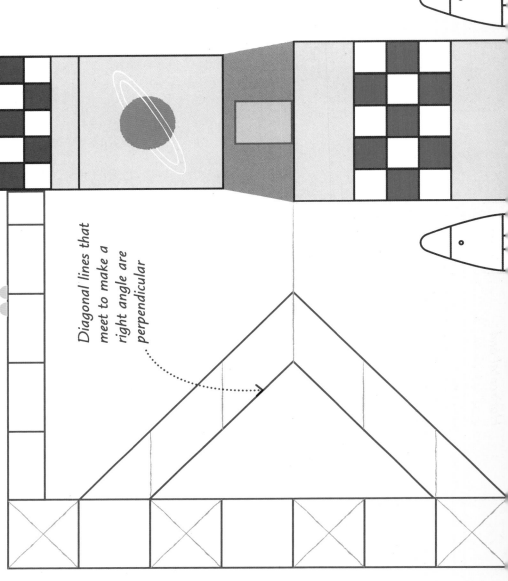

Diagonal lines that meet to make a right angle are perpendicular

1 Look at this picture of a rocket on a launch pad. You can see horizontal, vertical, and diagonal lines. Some of the lines are perpendicular.

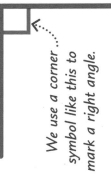

We use a corner symbol like this to mark a right angle.

2 When horizontal and vertical lines like these meet, we say they are perpendicular to each other. We call the point where they meet a right angle.

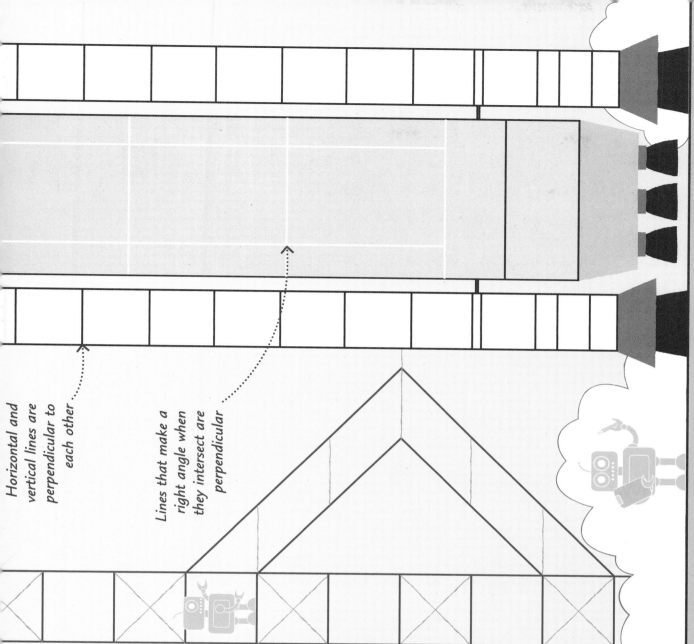

Horizontal and vertical lines are perpendicular to each other

Lines that make a right angle when they intersect are perpendicular

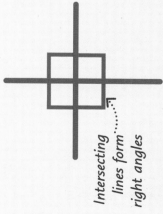

These diagonal lines meet to form a right angle

3 Any two lines that meet and make a right angle are perpendicular to each other. Perpendicular lines don't have to be horizontal and vertical.

Intersecting lines form right angles

4 Two lines are also perpendicular when they cross each other, or intersect, at a right angle.

5 Can you find more examples of all three kinds of perpendicular lines in the picture?

2D shapes

2D shapes are flat, like the shapes we draw on paper or on a computer screen. 2D is short for two-dimensional, because the shapes have length and height, or length and width, but no thickness.

Polygons and non-polygons

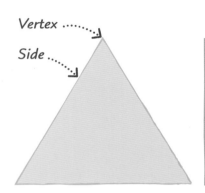

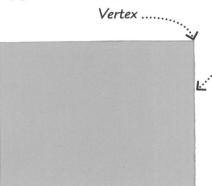

Vertex ········

Side ···

Vertex ········

Side

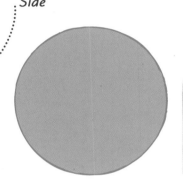

1 Polygons
Polygons are straight-sided shapes made of three or more sides and angles. Angles are made by two lines meeting at a point called a vertex.

2 Non-polygons
Other 2D shapes can be made from curved lines, like this circle, or by a combination of straight and curved lines, like the shape next to it.

Describing a polygon

We usually mark the sides of a polygon with dashes to show which sides are the same length as each other.

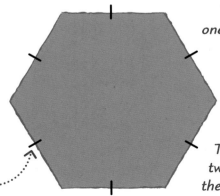

All sides are marked with one dash to show they are equal

The sides with one dash are the same length

The sides with two dashes are the same length

1 To show that all the sides are the same length, each side of this six-sided shape (hexagon) is marked with a single dash.

2 This hexagon has three sets of side of the same length. The first pair is marked with one dash, the second pair with two dashes, and the third pair with three dashes.

Regular and irregular polygons

A polygon is a 2D shape made of straight sides. Regular polygons have sides that are all the same length and angles of equal size. Irregular polygons have sides of different lengths and different-sized angles.

1 Triangle
A regular triangle has a special name – an equilateral triangle. Different irregular triangles also have special names. Find out more on page 215.

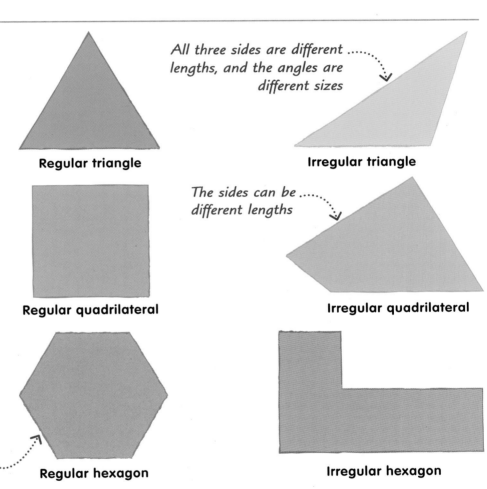

All three sides are different lengths, and the angles are different sizes

Regular triangle

Irregular triangle

2 Quadrilateral
Quadrilaterals are four-sided shapes. A regular quadrilateral is called a square.

The sides can be different lengths

Regular quadrilateral

Irregular quadrilateral

3 Hexagon
A six-sided polygon is called a hexagon.

All six sides are the same length, and the angles are equal

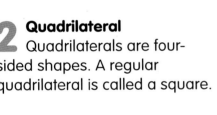

Regular hexagon

Irregular hexagon

TRY IT OUT

Odd one out

Only one of these five-sided polygons is a regular polygon, with sides of the same length and equal-sized angles. Can you spot it?

Answer on page 320

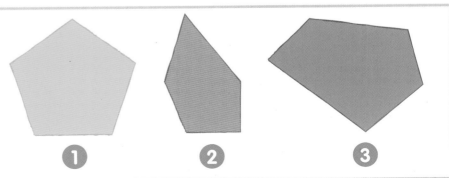

1   2   3

Triangles

A triangle is a type of polygon. It has three sides, three vertices, and three angles.

A triangle is a polygon with three straight sides and three angles.

Parts of a triangle

In geometry, we give special names to different parts of a triangle.

1 Side
The three straight lines that make up the triangle are called sides.

2 Vertex
A corner of a triangle, where two lines meet, is a vertex. The word for more than one vertex is vertices.

3 Base and apex
The base is the side that the triangle "rests" on. The apex is the vertex at the top of the triangle, opposite the base.

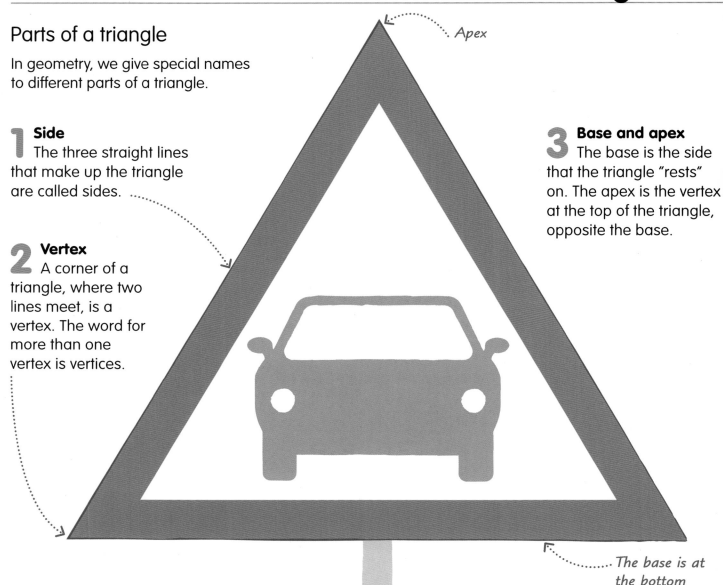

Apex

The base is at the bottom

Congruent triangles

Two or more triangles that have sides the same length and angles the same size are called congruent triangles. These triangles face different directions but are still congruent.

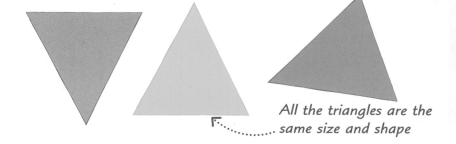

All the triangles are the same size and shape

ypes of triangles

Ve give triangles different names depending on the lengths of their ides and the sizes of their angles. On pages 240-41, you'll find out nore about the angles in triangles.

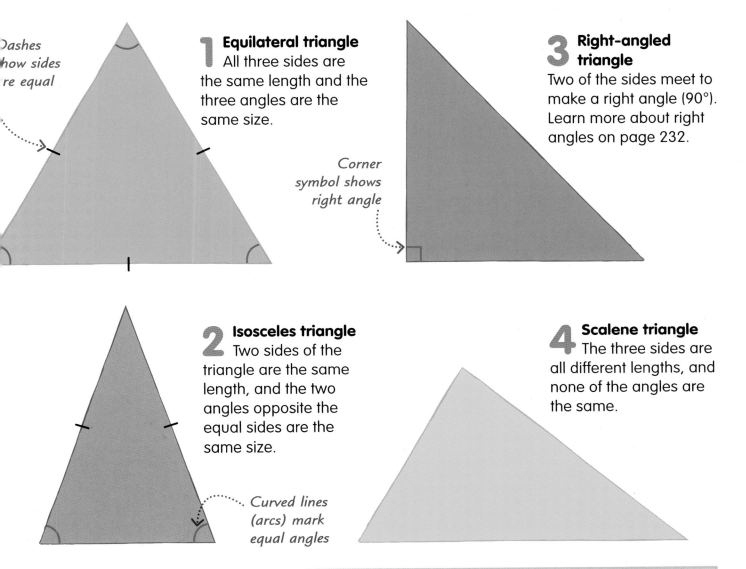

Dashes how sides re equal

1 Equilateral triangle
All three sides are the same length and the three angles are the same size.

Corner symbol shows right angle

3 Right-angled triangle
Two of the sides meet to make a right angle (90°). Learn more about right angles on page 232.

2 Isosceles triangle
Two sides of the triangle are the same length, and the two angles opposite the equal sides are the same size.

Curved lines (arcs) mark equal angles

4 Scalene triangle
The three sides are all different lengths, and none of the angles are the same.

TRY IT OUT

Triangle test

This picture contains different kinds of triangles. Can you spot an equilateral, an isosceles, a scalene, and a right-angled triangle?

Answers on page 320

Quadrilaterals

A quadrilateral is a polygon with four straight sides, four vertices, and four angles. "Quad" comes from the Latin word for "four".

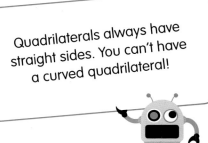

Quadrilaterals always have straight sides. You can't have a curved quadrilateral!

Types of quadrilaterals

Here are some of the most common quadrilaterals.

Opposite sides are marked with dashes to show they are the same length

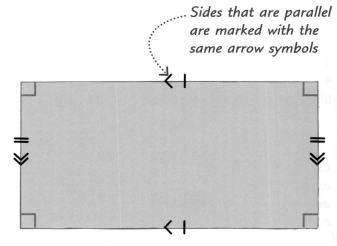

Sides that are parallel are marked with the same arrow symbols

1 Parallelogram
A parallelogram has two sets of parallel sides. Its opposite sides and opposite angles are equal.

2 Rectangle
The opposite sides of a rectangle are the same length and are parallel to each other. Each of its four angles is a right angle.

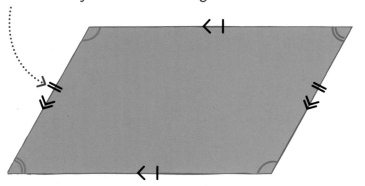

Angles that are equal are marked with curved lines, called arcs

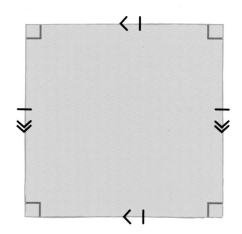

3 Rhombus
A rhombus has four sides of equal length. Its opposite sides are parallel, and its opposite angles are equal.

4 Square
A square has four sides of equal length. Each of its four angles is a right angle. The opposite sides of a square are parallel.

GEOMETRY • **QUADRILATERALS**

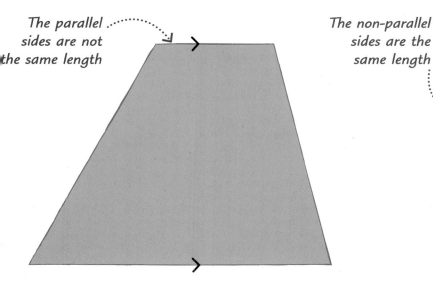

The parallel sides are not the same length

The non-parallel sides are the same length

5 Trapezium
A trapezium has one pair of parallel sides. It is also called a trapezoid.

6 Isosceles trapezium
This shape is like a normal trapezium, except that the non-parallel sides are the same length.

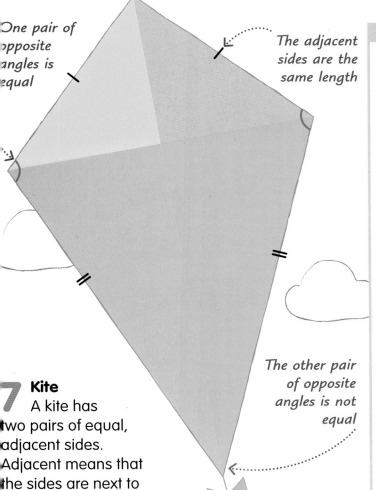

One pair of opposite angles is equal

The adjacent sides are the same length

The other pair of opposite angles is not equal

7 Kite
A kite has two pairs of equal, adjacent sides. Adjacent means that the sides are next to each other.

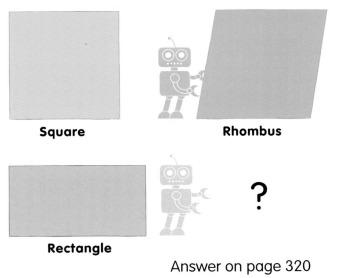

TRY IT OUT

Skewed shapes

Look at the square and the rhombus below. The rhombus looks like a skewed version of the square, as if it has been pushed sideways. Now look at the rectangle. If you skewed it in the same way, what shape would you get?

Square   **Rhombus**

Rectangle   **?**

Answer on page 320

Naming polygons

Polygons are named for the number of sides and angles they have. Most polygons' names come from the Greek words for different numbers. Here are some of the most common polygons.

> In a polygon, the number of its sides is always the same as the number of its angles.

3

3 sides and 3 angles

Regular triangle

Irregular triangle

6

6 sides and 6 angles

Regular hexagon

Irregular hexagon

7

7 sides and 7 angles

Regular heptagon

Irregular heptagon

REAL WORLD MATHS

Honey in hexagons

To store the honey they make, some bees build honeycombs from wax made inside their bodies. Honeycomb cells are regular hexagons that fit together perfectly to make a strong, space-saving storage unit for the honey.

10

10 sides and 10 angles

Regular decagon

Irregular decagon

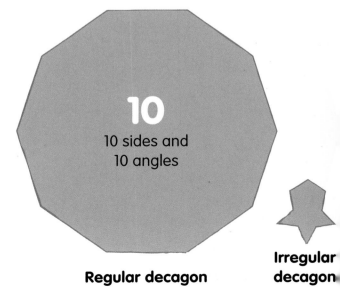

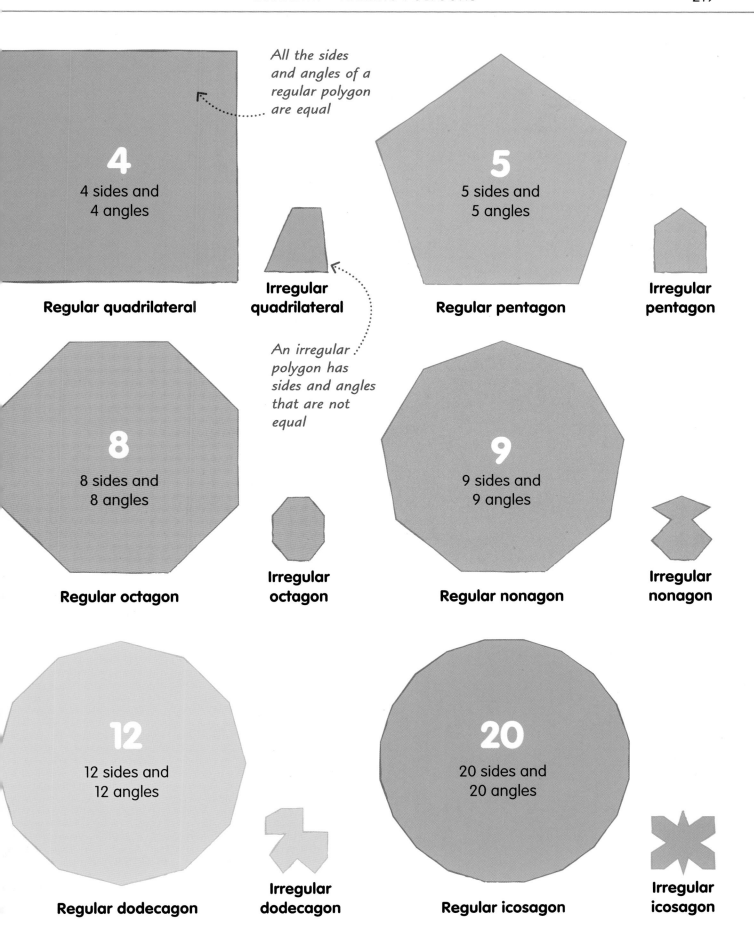

All the sides and angles of a regular polygon are equal

4
4 sides and
4 angles

Regular quadrilateral

Irregular quadrilateral

An irregular polygon has sides and angles that are not equal

5
5 sides and
5 angles

Regular pentagon

Irregular pentagon

8
8 sides and
8 angles

Regular octagon

Irregular octagon

9
9 sides and
9 angles

Regular nonagon

Irregular nonagon

12
12 sides and
12 angles

Regular dodecagon

Irregular dodecagon

20
20 sides and
20 angles

Regular icosagon

Irregular icosagon

Circles

A circle is a 2D shape, made from a curved line that goes all the way round a point at the centre. Every point on the line is the same distance from the centre.

The distance from the centre to any point on a circle's circumference is always the same.

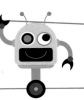

Parts of a circle

This drawing shows the most important parts of a circle. Some of these parts have special names that we don't use for other 2D shapes.

Arc

Circumference

Segment

Area

Diameter

Chord

Centre

Sector

Radius

Tangent

1 Circumference
The distance all the way round the circle. It's the circle's perimeter.

2 Radius
A straight line from the centre of the circle to the circumference. The plural of "radius" is "radii".

The diameter divides a circle in half

3 Diameter
A straight line from one side of the circle to the other, going through the centre. The diameter is twice the length of the radius.

4 Arc
Any part of the circle's circumference is called an arc.

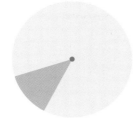

5 Sector
A slice of the circle formed by two radii and an arc.

6 Area
The amount of space inside the circle's circumference.

7 Chord
A line between two points on the circle's circumference that doesn't go through the centre.

8 Segment
The space between a chord and an arc.

9 Tangent
A straight line that touches the circumference at one point.

Use a ruler to measure the diameter

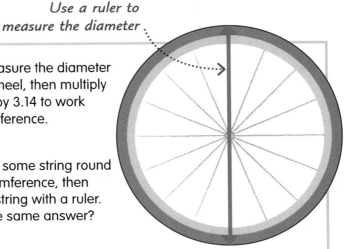

TRY IT OUT

Measure the circumference

A ruler won't help us measure a circle's circumference – it can't measure curves! But luckily, we can find the circumference of any circle if we multiply its diameter by 3.14.

Answer on page 320

1 First, measure the diameter of this wheel, then multiply the diameter by 3.14 to work out the circumference.

2 Now put some string round the circumference, then measure the string with a ruler. Do you get the same answer?

3D shapes

Three-dimensional, or 3D shapes, are shapes that have length, width, and height. A 3D object can be solid, like a lump of rock, or hollow, like a football.

All 3D shapes have three dimensions: length, width, and height. A 2D shape only has length and width or length and height.

1 Look at this picture of a greenhouse. It's made up of flat surfaces, joins, and corners. In geometry, these are called faces, edges, and vertices.

An edge is formed where faces meet

A vertex is formed where edges meet

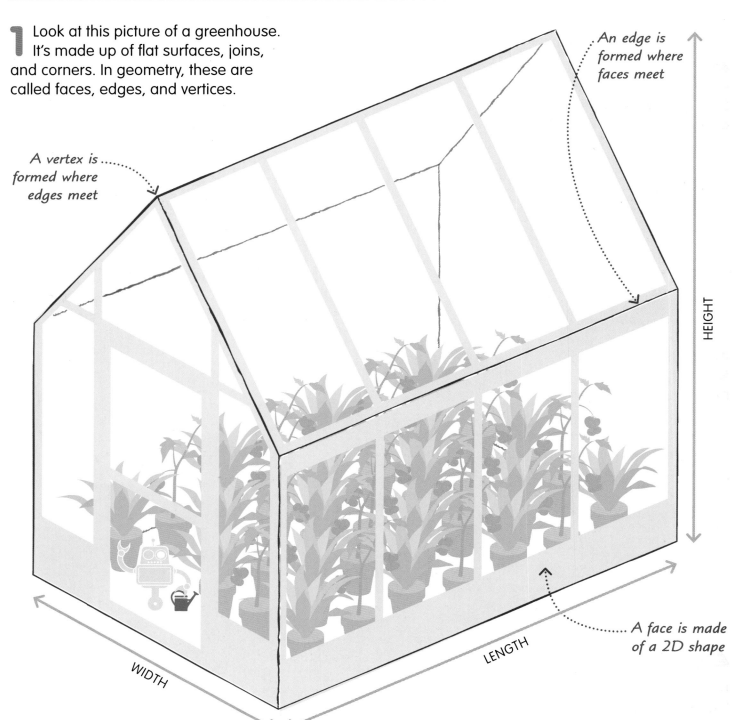

HEIGHT

WIDTH

LENGTH

A face is made of a 2D shape

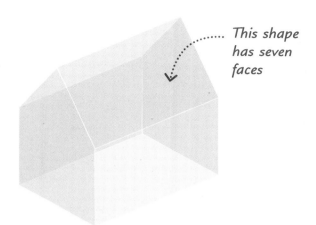

...... *This shape has seven faces*

Face
2 The surface of a 3D object is made of 2D shapes called faces. Faces can be flat or curved.

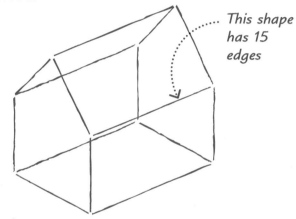

...... *This shape has 15 edges*

Edge
3 An edge is formed when two or more faces of a 3D shape meet.

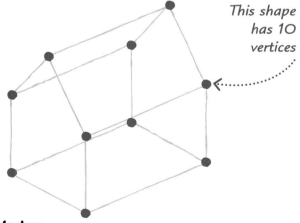

This shape has 10 vertices

Vertex
4 The point where two or more edges meet is called a vertex. The plural of "vertex" is "vertices".

Find the faces
Can you count the number of faces, edges, and vertices on this 3D shape?

Answers on page 320

It's a 3D world
Anything that has length, width, and height is 3D. Even a thin object, like a sheet of paper that's less than 1mm thick, has some height, so it's 3D, too. A complicated shape, like this plant in a pot, is also 3D, even though it's tricky to measure its different dimensions.

Types of 3D shape

3D objects can be any shape or size, but there are some that you will come across often in geometry. Let's take a closer look at some of the most common 3D shapes.

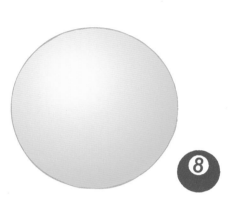

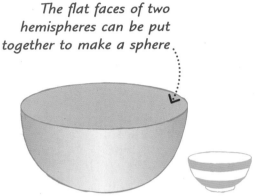

The flat faces of two hemispheres can be put together to make a sphere

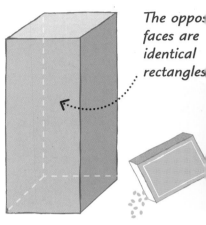

The oppos. faces are identical rectangles

1 Sphere
A sphere is a round solid. It has one surface and no edges or vertices. Every point on the surface is the same distance from the sphere's centre.

2 Hemisphere
A hemisphere is the name for half a sphere. It has one flat surface and one curved face.

3 Cuboid
A cuboid is a box-like shape with six faces, eight vertices, and 12 edges. Its opposite faces are identical.

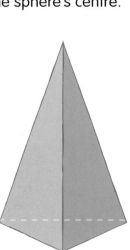

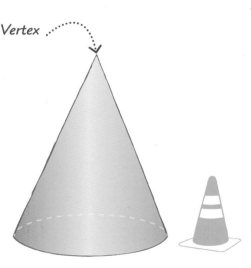

Vertex

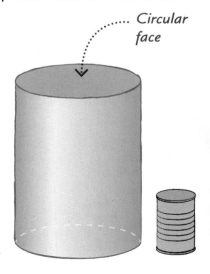

Circular face

5 Triangular-based pyramid
A triangular-based pyramid is also called a tetrahedron. It has four faces, four vertices, and six edges. It's unusual to see this kind of pyramid in the real world.

6 Cone
A cone has a circular base and a curved surface, which ends at a point directly above the centre of its base.

7 Cylinder
A cylinder has two identical circular ends joined by one curved surface.

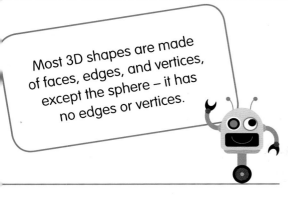

Most 3D shapes are made of faces, edges, and vertices, except the sphere – it has no edges or vertices.

Regular polyhedrons

A regular polyhedron is a 3D shape with faces that are regular polygons of the same shape and size. In geometry, there are only five regular polyhedrons. They are called the Platonic solids, after the Ancient Greek mathematician Plato.

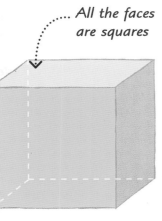

...... *All the faces are squares*

Cube

A cube is a special kind of cuboid. also has six faces, eight vertices, and 2 edges, but all its edges are the same ngth and all its faces are square.

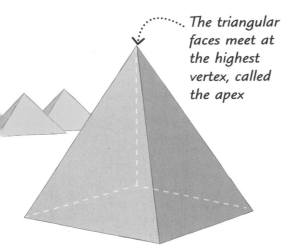

.... *The triangular faces meet at the highest vertex, called the apex*

Square-based pyramid

A square-based pyramid sits on a quare face. The other faces are triangles. has five vertices and eight edges.

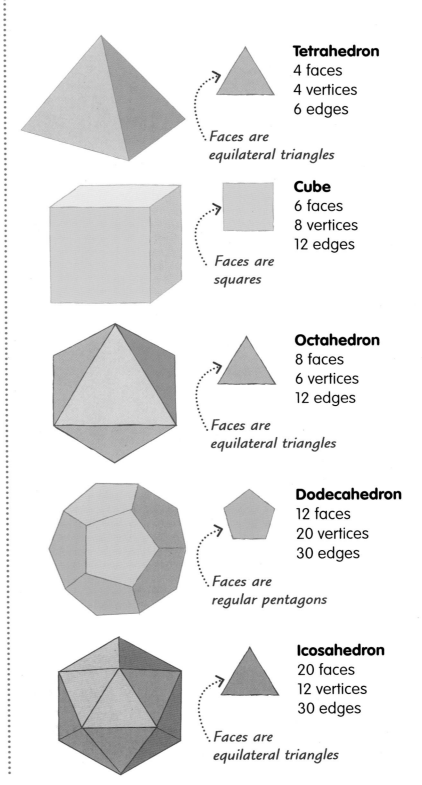

Tetrahedron
4 faces
4 vertices
6 edges

`. *Faces are equilateral triangles*

Cube
6 faces
8 vertices
12 edges

`. *Faces are squares*

Octahedron
8 faces
6 vertices
12 edges

`. *Faces are equilateral triangles*

Dodecahedron
12 faces
20 vertices
30 edges

`. *Faces are regular pentagons*

Icosahedron
20 faces
12 vertices
30 edges

`. *Faces are equilateral triangles*

Prisms

A prism is a special kind of 3D shape. It is a polyhedron, which means that all its faces are flat. Its two ends are also the same shape and size, and they are parallel to each other.

A prism is the same size and shape all the way along its length.

Finding prisms

Look at this picture of a campsite. We've pointed out some prisms, but can you spot them all? You should be able to find eight.

The tent shape's ends are parallel triangles, so we call it a triangular prism

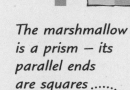
The marshmallow is a prism — its parallel ends are squares

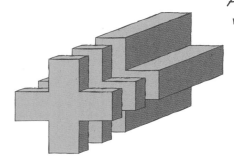

Cross sections

If you cut through a prism parallel to one of its ends, the new face you make is called a cross section. It will be the same shape and size as the original flat face.

All cross sections will be the same size and shape

Types of prism

There are many prisms in geometry. Here are some of the most common.

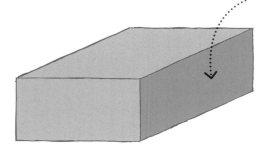

..... The sides of a prism are made of parallelograms

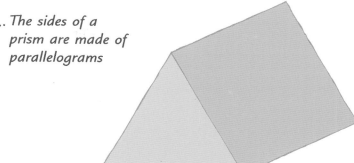

1 Cuboid
A cuboid is a prism. The opposite ends are rectangles, so we also call it a rectangular prism.

2 Triangular prism
A triangular prism, like the tent, has ends that are triangles.

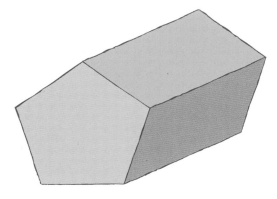

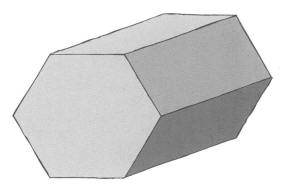

3 Pentagonal prism
A pentagonal prism has a pentagon at each end and five rectangular sides.

4 Hexagonal prism
A hexagonal prism's parallel ends are hexagons – six-sided polygons.

TRY IT OUT

Spot the non-prism

Which of these shapes is not a prism? Check to see if it has parallel faces at either end. Also, if you sliced through the shape, parallel to the end faces, would all the cross sections be the same?

Answer on page 320

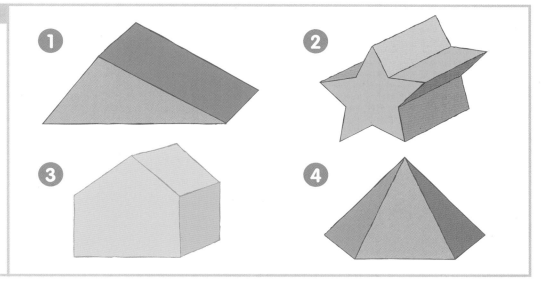

Nets

A net is a 2D shape that can be cut out, folded, and stuck together to make a 3D shape. Some 3D shapes, such as the cube on this page, can be made from many different nets.

A net is what a 3D shape looks like when it's opened out flat.

Net of a cube

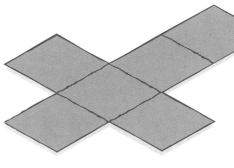

1 This shape, made of six squares, can be folded to make a cube. In geometry, we say the shape is a net of a cube.

Sharp creases help form the shape ....

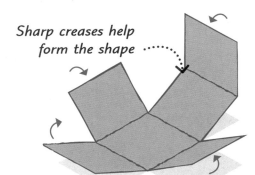

2 The shape is creased along the lines dividing the squares. When the lines are folded, they will form the edges of the cube.

The end square forms the lid ....

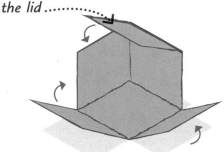

3 The squares round the central square will be the cube's sides. The square furthest from the centre square will be the lid.

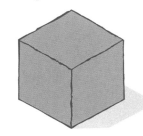

4 The flat net has now been turned into a cube.

TRY IT OUT

Find more nets

Here are three more nets of a cube. There are actually 11 different nets for a cube – can you work out any others?

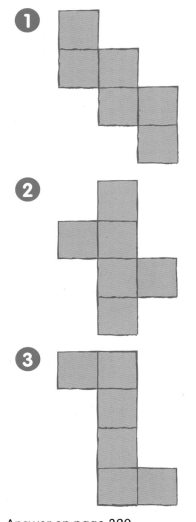

1

2

3

Answer on page 320

Nets for other 3D shapes

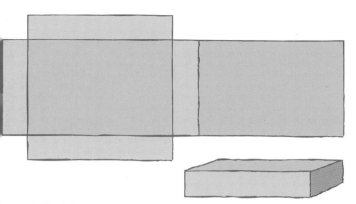

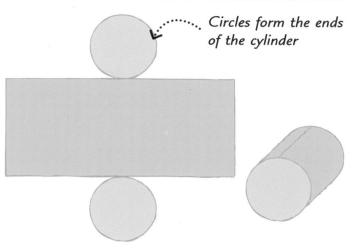

Circles form the ends of the cylinder

1 Cuboid
The net of a cuboid is made of six rectangles of three different sizes.

2 Cylinder
A cylinder's net is formed from just two circles and a rectangle.

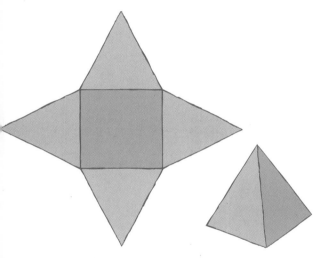

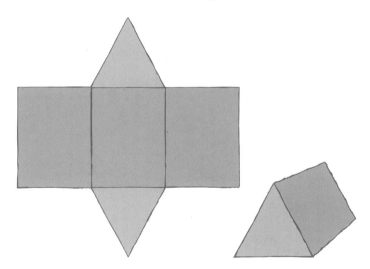

3 Square-based pyramid
One square and four triangles form the net of a square-based pyramid.

4 Triangular prism
A triangular prism is made from a net of three rectangles and two triangles.

REAL WORLD MATHS

Boxes need tabs

When we draw a net for a real 3D shape, we usually include tabs. Tabs are flaps added to some of the shape's sides so that we can stick the box together more easily. If you take an empty cereal box apart, you'll see the tabs that have been glued to some of the panels to form the box.

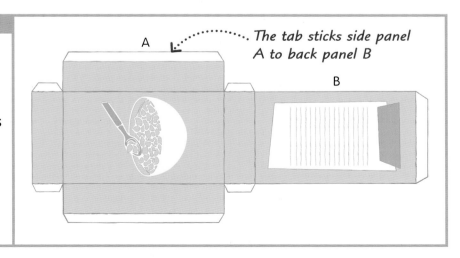

A

B

The tab sticks side panel A to back panel B

Angles

An angle is a measure of an amount of turn, or rotation, from one direction to another. It is also the difference in direction between two lines meeting at a point.

> An angle is a measure of the amount that something has turned around a fixed point.

1 Let's look at some lines turning round a centre point. As they turn, they create angles.

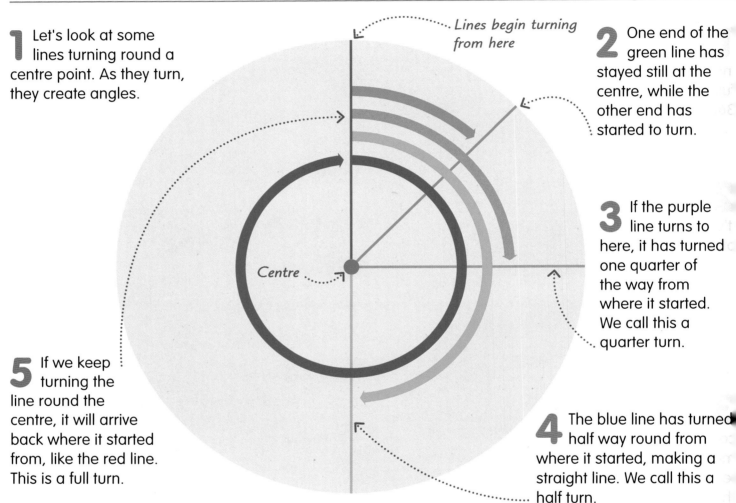

Lines begin turning from here

Centre

2 One end of the green line has stayed still at the centre, while the other end has started to turn.

3 If the purple line turns to here, it has turned one quarter of the way from where it started. We call this a quarter turn.

4 The blue line has turned half way round from where it started, making a straight line. We call this a half turn.

5 If we keep turning the line round the centre, it will arrive back where it started from, like the red line. This is a full turn.

Describing angles

An angle is made of three parts: two lines, called arms, and a vertex, where the arms meet. We show the angle by drawing a curved line, or arc, between the arms. The size of the angle is written inside or next to the arc.

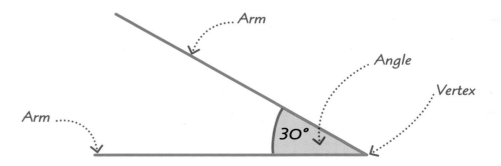

Arm

Angle

Vertex

Arm

30°

Degrees

We use units called degrees to precisely describe an amount of turn, which is how we measure the size of an angle. The symbol for degrees is a small circle, like this: °.

1 Here is a full turn divided into degrees. A full turn always has 360 degrees.

2 This is one degree (1°). It's equal to 1/360 of a full turn.

3 This shows ten degrees (10°). We can see that the angle made by this turn is ten times larger than the 1° angle.

4 This shows an angle of 100 degrees (100°).

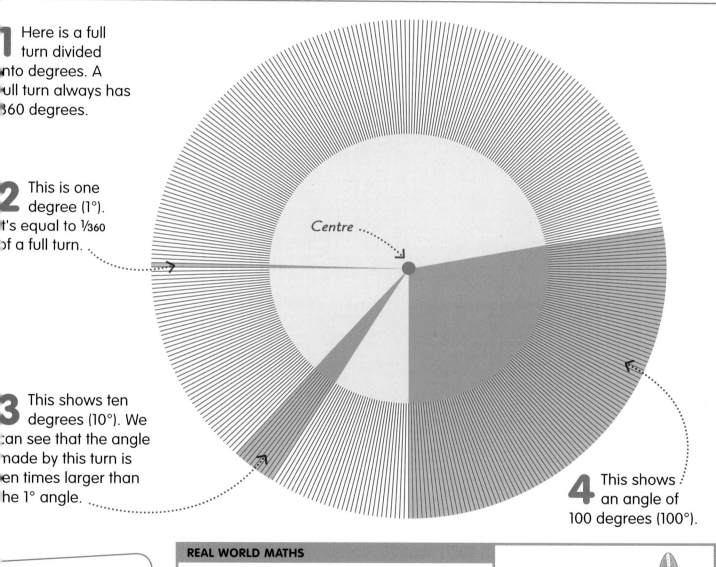

Centre

There are 360 degrees in a full turn.

REAL WORLD MATHS

Why 360 degrees?

One theory to explain why there are 360° in a full turn is that ancient Babylonian astronomers divided a full turn into 360 parts, because their year was 360 days long.

DAY 360

Right angles

Right angles are important angles in geometry. In fact, they are so important they get their own special symbol!

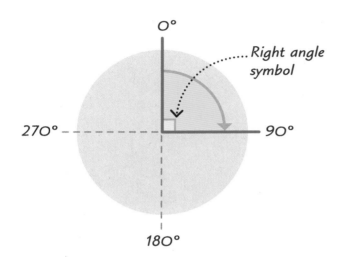

1 A quarter turn like this is 90°. We call it a right angle. When we mark a right angle, we make a corner symbol, like this: ⌐. We don't have to write "90°" next to the symbol.

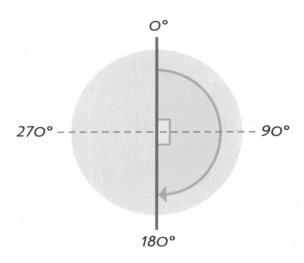

2 A half turn is 180°. It's also called a straight angle, because it makes a straight line. You can also think of a straight angle as two right angles.

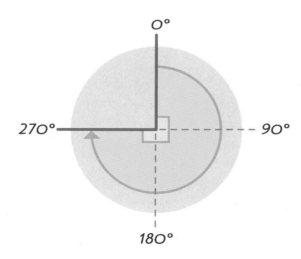

3 A three-quarter turn is 270°. It's made up of three right angles.

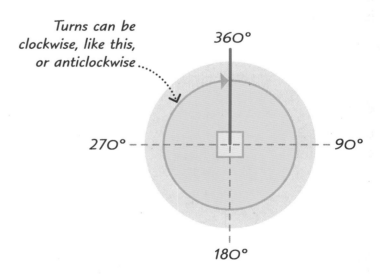

Turns can be clockwise, like this, or anticlockwise

4 A full turn is all the way round to where the line started, which is 360°. A full turn is made up of four right angles.

Types of angle

As well as the right angle, there are other important kinds of angle that we name according to their size.

1 Acute angle
When an angle is less than 90°, we call it an acute angle.

The arm turns anticlockwise to make a 45° angle

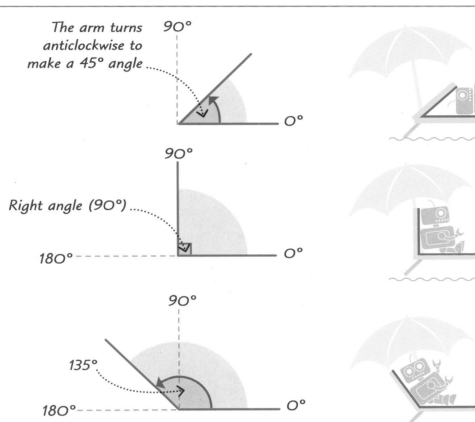

2 Right angle
A quarter turn is exactly 90°. We call it a right angle.

Right angle (90°)

3 Obtuse angle
An angle that's more than 90° but less than 180° is an obtuse angle.

4 Straight angle
An angle of exactly 180° is called a straight angle.

5 Reflex angle
An angle that's between 180° and 360° is called a reflex angle.

Angles on a straight line

Sometimes, simple rules can help us work out unknown angles.
One of these rules is about the angles that make up a straight line.

Angles on a straight line always add up to 180°.

1 If we rotated a line halfway round from where it started, the line would turn 180° and it would make a straight line.

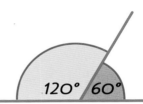

2 Imagine that your line made a stop on the way to the half turn, creating an extra line. The two angles made by the new line add up to 180°

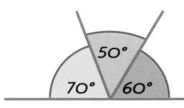

3 No matter how many angles you create on a straight line, they will add up to 180°, as long as all the lines start from the same point.

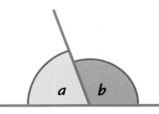

4 If the angles on a straight line are called a and b, we can write this rule as a formula:

$$a + b = 180°$$

Finding a missing angle on a straight line

1 Let's use the rule we've just learned to find the missing angle on this straight line.

2 We know that the three angles on the straight line add up to 180°.

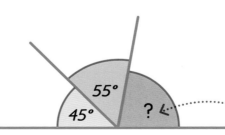

3 We know that one angle is 45° and the other is 55°. Let's add the angles together:
45° + 55° = 100°

4 Now let's subtract that total from 180°:
180° − 100° = 80°

5 So the missing angle is 80°.

Angles at a point

Another rule of geometry is that angles that meet at a point always add up to 360°. This rule helps us work out missing angles when they surround a point.

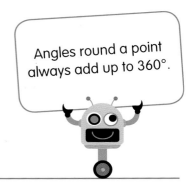

Angles round a point always add up to 360°.

1 We know that if we turn a line all the way round to where it started, it makes a full turn, which is 360°.

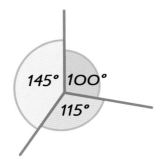

2 Imagine that the line stops on its way to making a full turn, creating new lines that meet at the same point. The angles formed all add up to 360°.

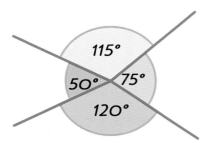

3 This time, there are four lines meeting at a point. But it doesn't matter how many lines there are – the angles will always add up to 360°.

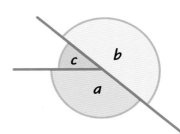

4 If the angles that meet at a point are called a, b, and c, we can write this rule as a formula:
$$a + b + c = 360°$$

Finding the missing angle round a point

1 Let's use the rule we've just learned to find the missing angle at this point.

2 We know that the three angles round this point add up to 360°.

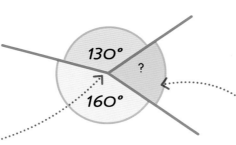

3 We also know that one angle is 160° and the other is 130°. Let's add these angles together:
160° + 130° = 290°

4 Now let's subtract that total from 360°: 360° − 290° = 70°

5 This means that the missing angle is 70°.

Opposite angles

When two straight lines cross, or intersect, they create two pairs of matching angles called opposite angles. We can use this information to work out angles we don't know.

> When two lines intersect, the angles directly opposite each other are always equal.

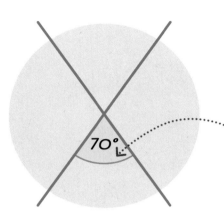

Use a protractor to measure the angle

1 Let's look at what's special about opposite angles. First, we draw two intersecting straight lines, then measure the bottom angle.

The opposite angles are coloured the same blue to show they are equal

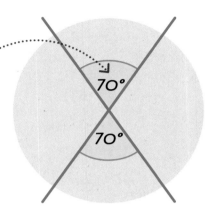

2 When we measure the top angle, we find it's the same as the bottom one. The angles opposite each other are equal.

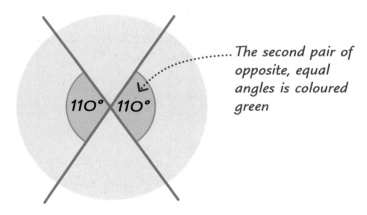

The second pair of opposite, equal angles is coloured green

3 Now let's look at the other pair of opposite angles. When we measure them, we find that they are also equal – they are both 110°.

4 If we call the angles a, b, c, and d, we can write what we know about opposite angles like this:

a = c   b = d

Finding missing angles

When two lines intersect, if we know the size of one angle, we can work out the sizes of all the others.

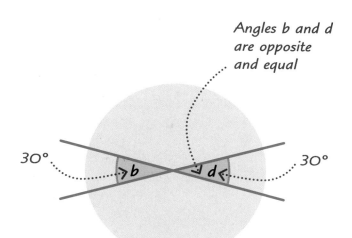

Angles b and d are opposite and equal

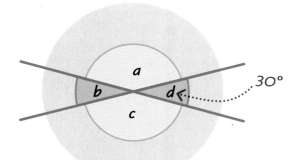

... 30°

1 These two lines intersect, creating two pairs of opposite angles. We know that angle d is 30°.

2 Angles b and d are opposite each other, so we know that angle b must be 30°, too.

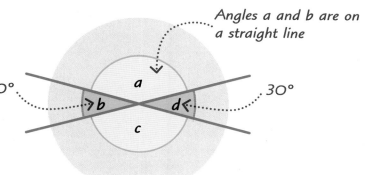

Angles a and b are on a straight line

30° ... 30°

3 We can use what we know about angles on a straight line to work out angle a. We know that a + b = 180°, so a must be 180° − 30°. So a = 150°.

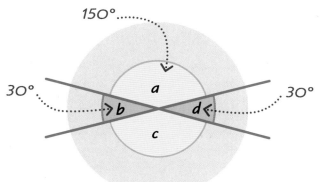

150° .....

30° ... 30°

4 Angles a and c are opposite, so we know that means they are equal. So c is 150°.

TRY IT OUT

Angles brainteaser

Can you work out these missing angles? Use what you know about the size of a right angle, the angles on a straight line, and that opposite angles are equal.

Answers on page 320

Add a and b to make d's opposite angle .....

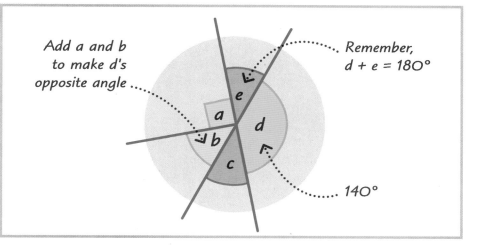

Remember, d + e = 180°

..... 140°

Using a protractor

We use a protractor to draw and measure angles accurately. Some protractors measure angles up to 180°, while others can measure angles up to 360°.

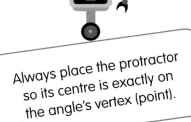

> Always place the protractor so its centre is exactly on the angle's vertex (point).

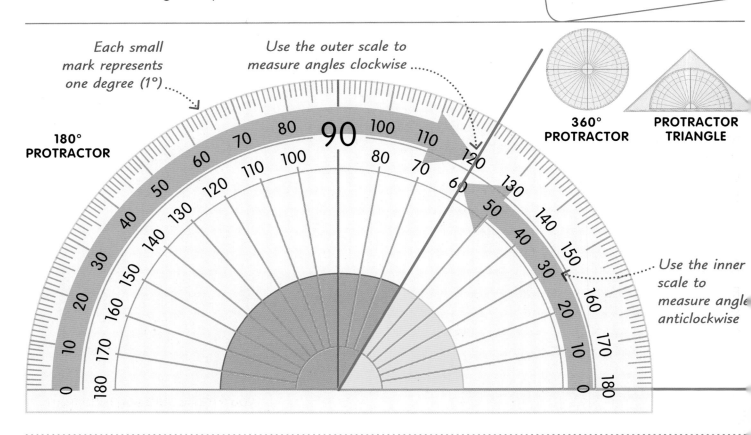

Each small mark represents one degree (1°)

Use the outer scale to measure angles clockwise

180° PROTRACTOR

360° PROTRACTOR

PROTRACTOR TRIANGLE

Use the inner scale to measure angle anticlockwise

Drawing angles

A protractor is essential if you need to draw an angle accurately.

Mark a point on the line

1 Here's how to draw a 75° angle. Draw a straight line with a pencil and ruler, and mark a point on it.

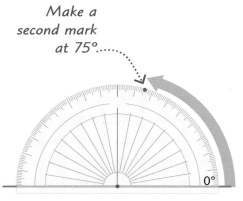

Make a second mark at 75°

2 Put the protractor's centre on the marked point. Read up from 0°, and make a second mark at 75°.

Draw a straight line between the two points

3 Use a ruler and pencil to draw a line between the two points, then label the angle.

75°

Measuring angles up to 180°

You can use a protractor to measure the angle formed by any two lines.

Put the centre of the protractor over the vertex

Use the inner scale to measure the smaller angle

Use the outer scale to measure the larger angle

Make the arms longer if they're not long enough to read

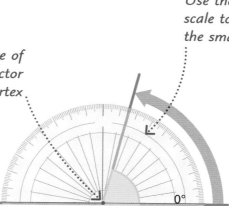

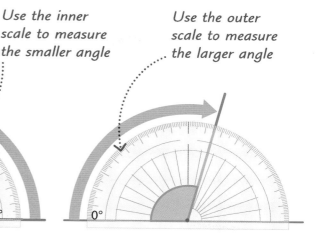

1 Use a ruler and pencil to extend the angle's arms if you need to. This makes it easier to read the angle.

2 Put the protractor along one arm of the angle. Take a reading from where the other arm crosses the protractor.

3 To measure the larger angle, read up from zero on the other side of the protractor.

Measuring reflex angles

Reflex angles are angles larger than 180°. We can use a semicircular protractor to measure a reflex angle if we combine our measurements with what we know about calculating angles.

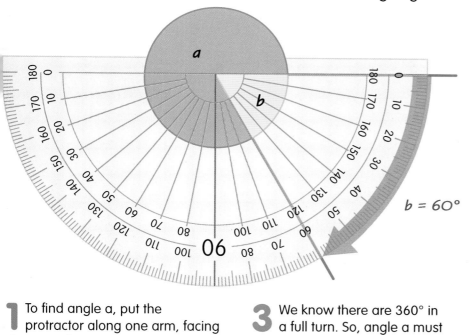

b = 60°

1 To find angle a, put the protractor along one arm, facing downwards.

2 When we measure angle b, we find that it's 60°.

3 We know there are 360° in a full turn. So, angle a must be 360° − 60°.

4 So, the answer is a = 300°.

Measure the angles

Practise your protractor skills by measuring these angles. It helps to estimate angles before measuring them – that way, you'll make sure you read from the correct scale.

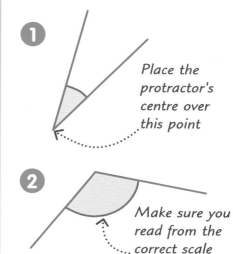

1 *Place the protractor's centre over this point*

2 *Make sure you read from the correct scale*

Answers on page 320

Angles inside triangles

We give names to triangles according to the lengths of their sides and the sizes of their angles. We learned about the sides of a triangle on page 214, so now let's have a closer look at its angles.

Strong shapes

Triangles are useful shapes for engineers as they are stable and hard to pull out of shape. This geodesic dome is made from triangular panels, which work together to carry weight evenly. This makes the structure light, but very strong.

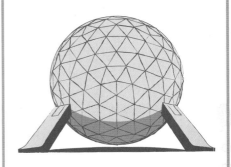

Types of triangles

Here are the triangles we see most often in geometry.

There are four kinds of triangle: equilateral, right-angled, isosceles, and scalene.

Equal angles are marked with arcs

60°

60°   60°

Equal sides are marked with a dash

The two angles that are not right angles can be the same or different

1 Equilateral triangle

An equilateral triangle is the more usual name for a regular triangle. Its three angles are all 60°. Its three sides are always the same length, too.

2 Right-angled triangle

A right-angled triangle contains one right angle, which is exactly 90°. The other two angles can be the same, or different, like this one. It can have two sides of the same length, or all three can be different.

3 Isosceles triangle

An isosceles triangle has two angles of equal size and two sides of the same length. The third angle can be any size.

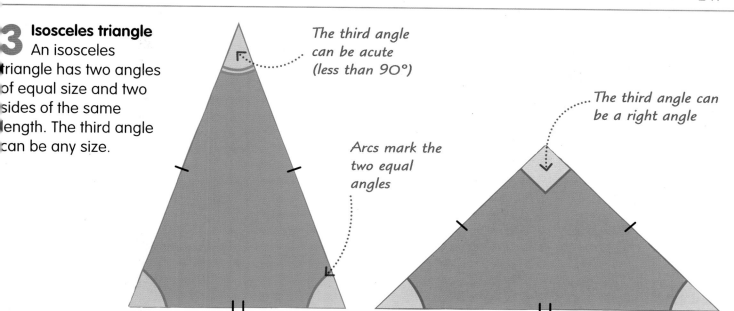

The third angle can be acute (less than 90°)

The third angle can be a right angle

Arcs mark the two equal angles

4 Scalene triangle

A scalene triangle has no equal sides, and all its angles are different. It can contain one right angle, or it can be made up of a combination of acute and obtuse angles.

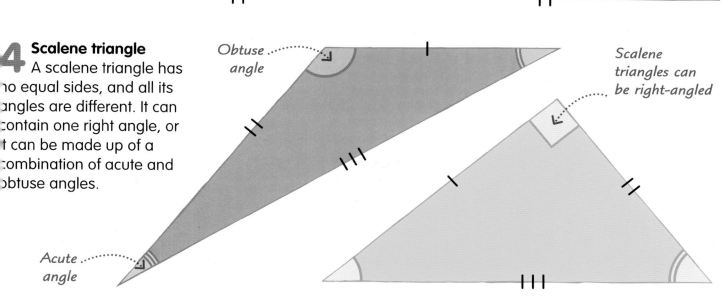

Obtuse angle

Acute angle

Scalene triangles can be right-angled

TRY IT OUT

Work out the angles

If you know what type of triangle you're looking at, you can sometimes work out all its angles, even if you only know one of them. See if you can work out the two missing angles here. The steps will help you if you get stuck.

Answer on page 320

1 This is an isosceles triangle, so we know that a and b are equal.

2 We know that a + b + c = 180°. Angle c is 40°, so if we take 40° away from 180°, the answer will be the same as a + b.

3 Now, if we divide that answer by two, we find the size of angles a and b.

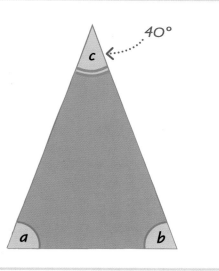

40°

Calculating angles inside triangles

The special thing about the angles inside a triangle is that they always add up to 180°. It doesn't matter whether the lengths of the sides and angles are the same or different – when we add the angles up, we always get the same answer.

1 Look at the three sails on this boat. Each one is a triangle, but all three triangles are different.

2 This triangle has angles of 60°, 30°, and 90°. Let's add them up:
60° + 30° + 90° = 180°

70°
40°
70°
30°

3 In this triangle, two of the angles are the same. When we add all three angles we get:
70° + 70° + 40° = 180°

40°
90°
90°
50°
60°

4 The third triangle is different again but we still get the same answer:
40° + 50° + 90° = 180°

Prove it!

One way to test that the angles inside a triangle add up to 180° is to take the three corners from a triangle and see how perfectly they fit along a straight line. We already know a straight line is 180°.

Tear each corner off

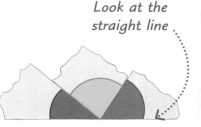

Look at the straight line

1 Cut a triangle out of paper. The sides and angles can be any size. Now tear off the three corners.

2 Rotate the three corners so that you can bring them together.

3 Make all three corners touch. Look how they form a straight line, which we know is 180°.

Finding a missing angle in a triangle

The rule we've just learned can be really useful, because if we know the size of two of the angles inside a triangle, we can work out the third.

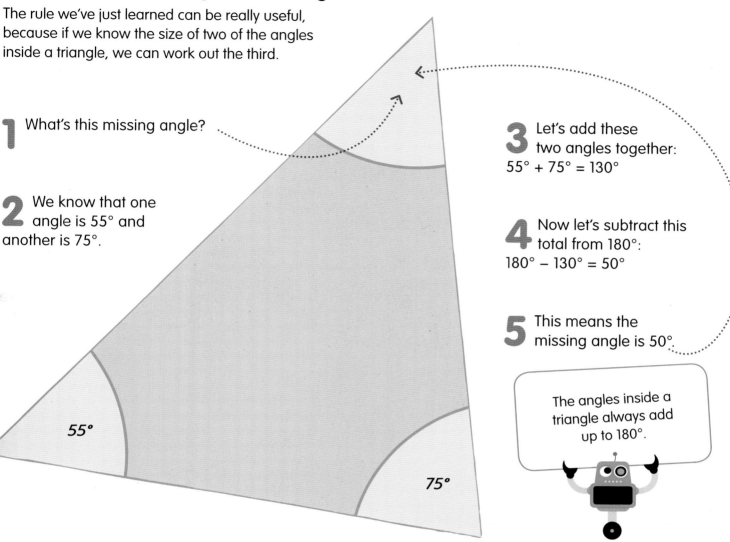

1 What's this missing angle?

2 We know that one angle is 55° and another is 75°.

3 Let's add these two angles together:
55° + 75° = 130°

4 Now let's subtract this total from 180°:
180° − 130° = 50°

5 This means the missing angle is 50°.

The angles inside a triangle always add up to 180°.

55°

75°

TRY IT OUT

Find the mystery angles

Now use the method we've just learned to find the missing angles in these triangles.

Answers on page 320

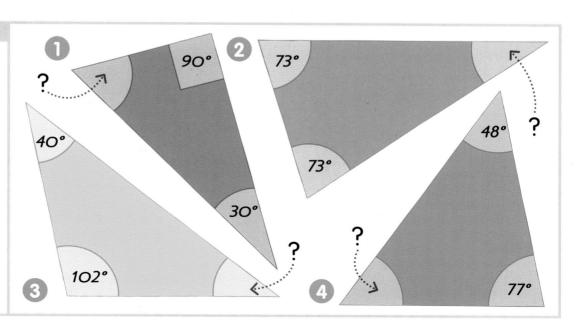

1 ?   90°   40°   30°

2 73°   73°   48°   ?

3 102°   ?

4 ?   77°

Angles inside quadrilaterals

Quadrilaterals have different names, depending on the properties of their sides and angles. We looked at a quadrilateral's sides on pages 216-17. Now let's have a closer look at its angles.

All quadrilaterals have four angles, four sides, and four vertices.

Types of quadrilateral

Quadrilaterals are polygons with four sides and four angles. Here are some of the quadrilaterals we see most often in geometry.

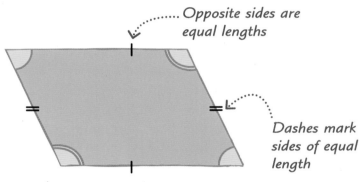

Opposite sides are equal lengths

Dashes mark sides of equal length

1 Parallelogram
A parallelogram has two pairs of equal angles, opposite to each other.

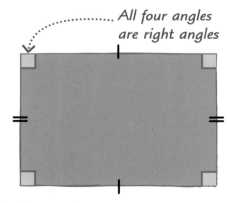

All four angles are right angles

2 Rectangle
A rectangle has four right angles and two pairs of equal, parallel sides.

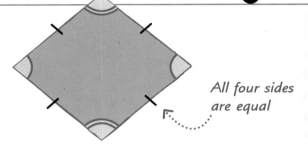

All four sides are equal

3 Rhombus
The opposite angles of a rhombus are equal. Another name for a rhombus is a diamond.

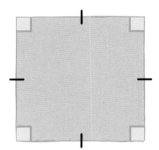

4 Square
A square is special kind of rectangle, with four right angles and four equal sides.

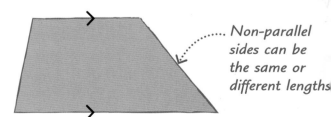

Non-parallel sides can be the same or different lengths

5 Trapezium
Two of a trapezium's angles are greater than 90°. It has one pair of parallel sides.

Calculating angles inside quadrilaterals

The angles inside a quadrilateral always add up to 360°.
There are two ways we can prove that this is true.

The angles inside a quadrilateral always add up to 360°.

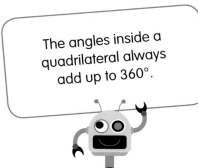

1 **Make triangles**
A quadrilateral can be split into two triangles, like this. We know that the angles in a triangle add up to 180°. That means the angles in a quadrilateral add up to 2 × 180°, which is 360°.

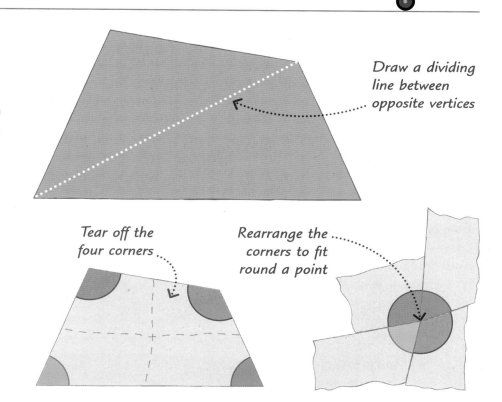

Draw a dividing line between opposite vertices

2 **Put the angles round a point**
You can tear the corners off a quadrilateral and arrange them round a point, like this. We know that angles round a point add up to 360°, so the quadrilateral's angles must add up to 360°, too.

Tear off the four corners

Rearrange the corners to fit round a point

Find the missing angle

So, now we know that the angles in a quadrilateral add up to 360°.
We can use this fact to work out missing angles in quadrilaterals.

1 Look at this shape. What's the missing angle?

2 We know that three of the angles are 75°, 95°, and 130°. Lets add them together:
75° + 95° + 130° = 300°

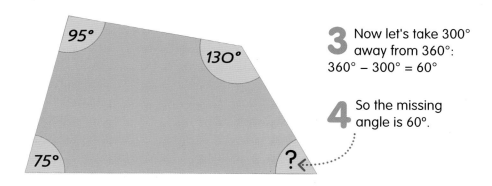

95°

130°

75°

?

3 Now let's take 300° away from 360°:
360° − 300° = 60°

4 So the missing angle is 60°.

Angles inside polygons

Polygons get their names from the number of their sides and angles. We learned about polygons' sides on pages 218-19. Now we're going to focus on their angles.

> The sum of the angles inside a polygon depends on how many sides it has.

More sides means bigger angles

All the angles in a regular polygon are the same size. So, if you know one angle, you know them all. Look at these polygons. You can see that the more sides a regular polygon has, the larger its angles become.

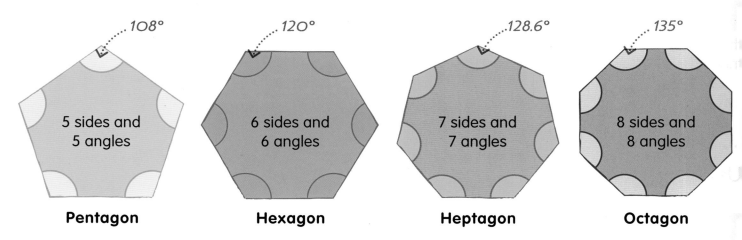

108°	120°	128.6°	135°
5 sides and 5 angles	6 sides and 6 angles	7 sides and 7 angles	8 sides and 8 angles
Pentagon	**Hexagon**	**Heptagon**	**Octagon**

Angles inside regular and irregular polygons

The angles inside polygons with the same number of sides always add up to the same amount. Let's look at the angles inside two different hexagons.

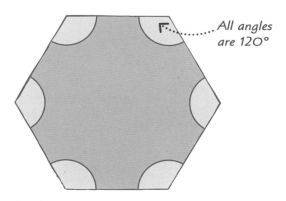

All angles are 120°

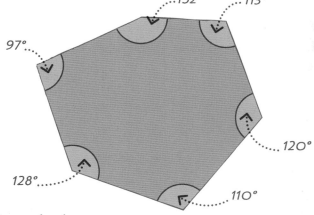

152°   113°

97°

120°

128°   110°

1 **Regular hexagon**
The angles inside this regular hexagon are all the same size. The six angles of 120° add up to a total of 720°.

2 **Irregular hexagon**
In this irregular hexagon, each angle is different. But when you add them up, the total is 720° – the same as for the regular hexagon.

Calculating the angles in a polygon

To find the sum of all the angles inside a polygon, we can either count the triangles it contains, or use a special formula.

Counting triangles

1 Look at this pentagon. You can see that we can divide the five-sided shape into three triangles.

A pentagon can be split into three triangles

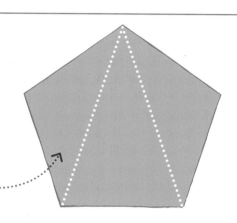

2 We know that the angles in a triangle add up to 180°. The pentagon is made of three triangles, so the angles add up to 3 × 180°, which is 540°

Using a formula

1 Here's a rule about the angles in polygons: the number of triangles a polygon can be divided into is always two fewer than the number of its sides.

2 Let's look at the pentagon again. It has five sides, which means it can be divided into three triangles.

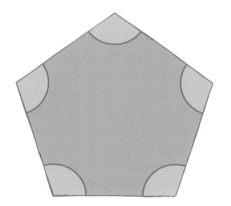

3 So, we can write the sum of the angles in a pentagon like this: (5 − 2) × 180°= 3 × 180° = 540°.

4 There's a formula that works for all polygons. If we call the number of sides n, then:

SUM OF ANGLES IN A POLYGON
= (n-2) × 180°

TRY IT OUT

Polygon poser

Combine what you've learned about angles inside a polygon to work out the seventh angle in this irregular heptagon. Remember, if you know how many sides a polygon has, you can work out the sum of its angles.

Answer on page 320

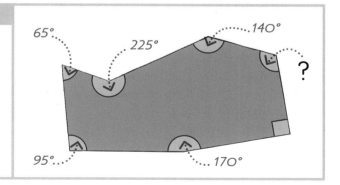

Coordinates

Coordinates help us to describe or find the position of a point or place on a map or grid. Coordinates come in pairs, to tell us how far along and up or down the point is.

In a pair of coordinates, the x coordinate always comes before the y coordinate.

Coordinate grids

The y axis is vertical

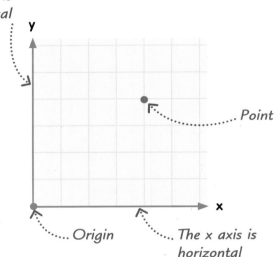

Point

Origin

The x axis is horizontal

1 This grid is called a coordinate grid. It's made up of horizontal and vertical lines that cross, or intersect, to make squares.

2 The two most important lines on the grid are the x axis and the y axis. We use them to help us describe the coordinates of points on the grid.

3 The x axis is always horizontal, and the y axis is vertical.

4 The point on the grid where the x and y axes intersect is called the origin.

Finding the coordinates of a point

The position of any point on a grid can be described by its coordinates.

The x coordinate of A is 2, which is two squares along

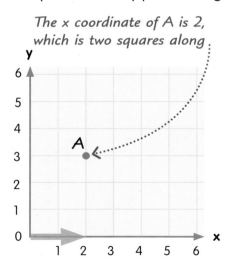

The y coordinate of A is 3, which is three squares up

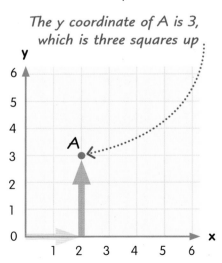

Coordinates are always in brackets

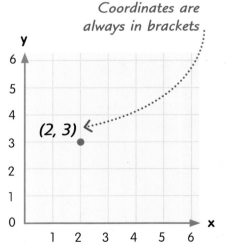

(2, 3)

1 To find the coordinates of A, first we count how many squares it is along the x axis. It is two squares along from the origin, so the x coordinate is 2.

2 Now we read up the y axis to count how many squares up it is to the point. It is three squares up from the origin, so we say that the y coordinate is 3.

3 We write the point's coordinates as (2, 3), which means two squares along and three up. We put coordinates in brackets.

Plotting points using coordinates

We can also use coordinates to place, or plot, points accurately onto a grid.

Marking a specific place on a grid is called plotting a point.

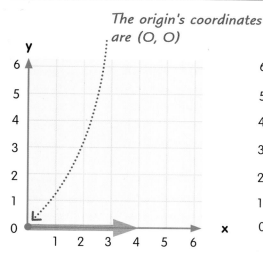

The origin's coordinates are (O, O)

1 To plot the coordinates (4, 2), we first count four squares along the x axis.

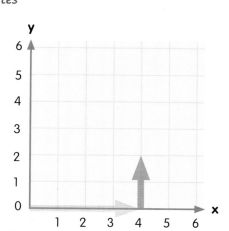

2 Next, we count two squares up the y axis.

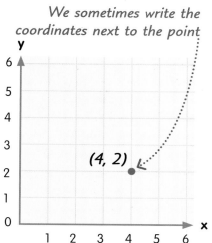

We sometimes write the coordinates next to the point

(4, 2)

3 Now we mark the point we have reached with a dot.

TRY IT OUT

Find the coordinates

Can you write down the coordinates of points A, B, C, and D? Remember the x coordinate is written first, then the y coordinate.

Answers on page 320

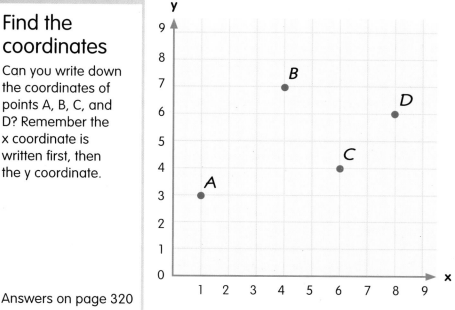

REAL WORLD MATHS

Grids and maps

One of the most common ways we use coordinates on a grid is to find locations on a map. Most maps are drawn with a coordinate grid.

Positive and negative coordinates

The x and y axes on a grid can go either side of zero, just as they do on a number line. On this kind of grid, a point's position is described with positive and negative coordinates.

Quadrants of a graph

When we extend the x and y axes of a grid beyond the origin, we create four different sections. These are called the first, second, third, and fourth quadrants.

Coordinates can be positive or negative, depending on the quadrant they are located in

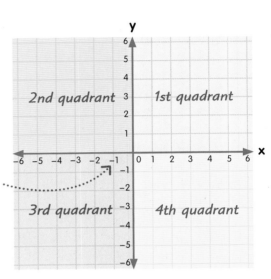

Plotting positive and negative coordinates

Points on a grid can have positive or negative coordinates, or a mixture of both, depending on which quadrant they are in.

1 In the first quadrant, both coordinates are made of positive numbers. Point A is two squares along the x axis and 4 squares up the y axis, so its coordinates are (2, 4).

2 In the second quadrant, point B is 2 squares behind the origin (0,0), so the x coordinate is −2. It's 3 squares up on the y axis, so point B's coordinates are (−2, 3).

3 In the third quadrant, point C is behind the origin on the x axis and below it on the y axis, so both coordinates are negative numbers. The coordinates are (−5, −1).

4 In the fourth quadrant, point D is 6 squares along the x axis and 3 down on the y axis. So, its coordinates are (6, −3).

Both coordinates are positive

The x coordinate is negative and the y coordinate is positive

Both coordinates are negative

The x coordinate is positive and the y coordinate is negative

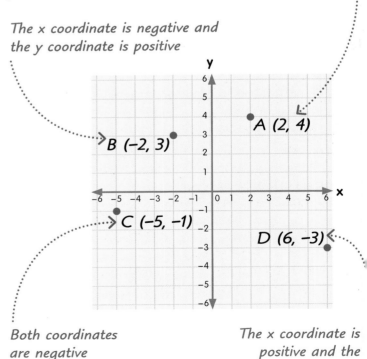

Using coordinates to draw a polygon

We can draw a polygon on a grid by plotting its coordinates, then joining the points with straight lines.

Remember, positive or negative numbers in coordinates tell us in which quadrant we will find a point.

How to plot and draw a polygon on a grid

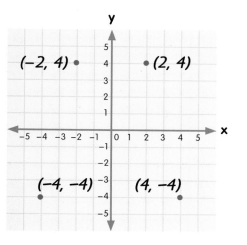

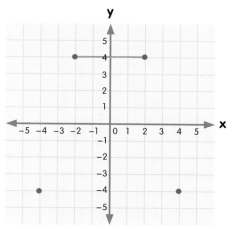

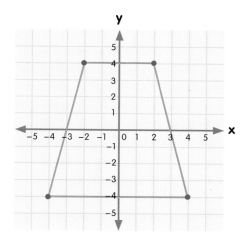

1 We start by plotting these four coordinates on the grid:
(2, 4); (−2, 4); (−4, −4); (4, −4).

2 Now we use a pencil and ruler to join up the first two points we plotted.

3 We carry on joining up the points until we have made a shape called a trapezium.

TRY IT OUT

Plotting posers

1 Can you work out the coordinates that make the points of this six-sided shape, called a hexagon?

2 If you plotted these coordinates on a grid and joined the points in order with straight lines, what shape would you draw?
(1, 0) (0, −2) (−2, −2) (−3, 0) (−1, 2)

Answers on page 320

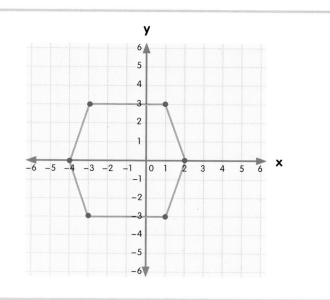

Position and direction

We can use a grid and coordinates to describe the positions of places on a map.

How to use coordinates on a map

Maps are often divided up by a square grid, so we can pinpoint the position of a place by giving its square's coordinates.

1 Every square on the map has a unique pair of coordinates that describe its position.

2 The first coordinate tells us how far along the grid to count horizontally. The second coordinate tells how many squares to count up vertically.

The vertical grid is marked with numbers

This square's coordinates are B2

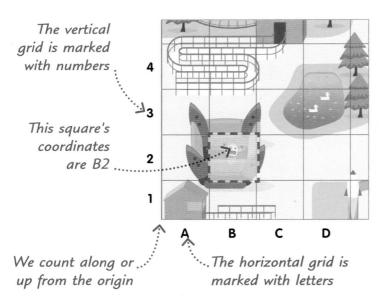

We count along or up from the origin

The horizontal grid is marked with letters

3 This map uses letters for the horizontal coordinates and numbers for the vertical coordinates. Often, maps use numbers for both the horizontal and vertical coordinates.

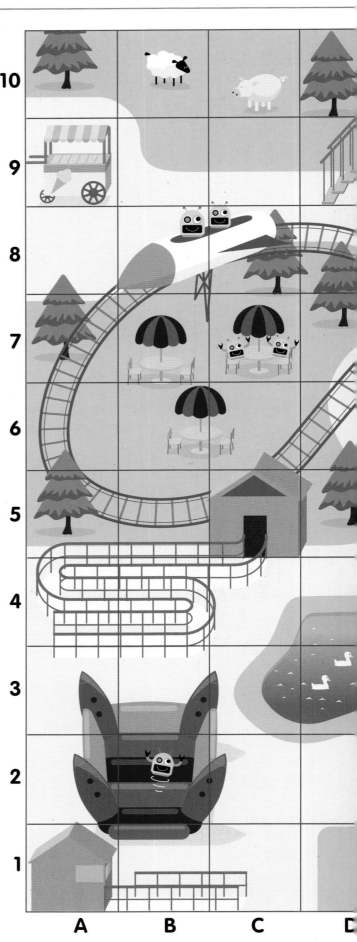

E   **F**   **G**   **H**   **I**   **J**

4 We can use map coordinates to find our way round Cybertown's theme park, Astro World. The sheep in the petting zoo is two squares along and 10 squares up. Its coordinates are B10.

5 The ducks in the pond are four squares along and three squares up. So, their coordinates are D3.

6 To find what's in square A9, we count one square to the right and nine squares up. The square contains the ice cream cart.

TRY IT OUT

Find the spot

See if you can navigate your way round the map by finding these things:

1 What can you find at square G10?

2 Now find H3. What's in the square?

3 Can you give the coordinates of the table with two robots sitting at it?

Answers on page 320

Compass directions

A compass is a tool we use for finding a location or to help us move in a particular direction. It has a pointer that always shows the direction of north.

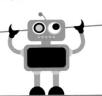

> The four cardinal compass points are: north (N), south (S), east (E), and west (W).

Points on a compass

Compass points show directions as angles measured clockwise from the direction north. We call these directions bearings.

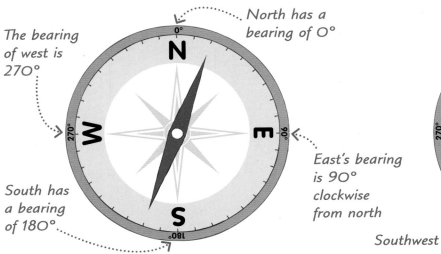

The bearing of west is 270°

North has a bearing of 0°

South has a bearing of 180°

East's bearing is 90° clockwise from north

Northwest is halfway between north and west

Northeast

Southwest

Southeast

1 The main compass points are: north (N), south (S), east (E), and west (W). We call them the cardinal points.

2 Halfway between the cardinal points are the ordinal points: northeast (NE), southeast (SE), southwest (SW), and northwest (NW).

Using a compass with a map

Most maps are printed with a north arrow. If we align north on the compass with north marked on a map, we can find the directions to other locations on the map. We can then use our compass to get from one place to the other.

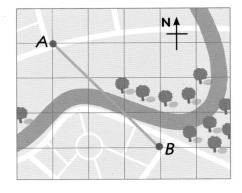

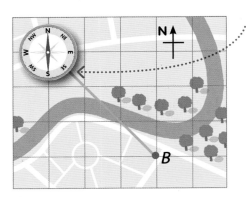

Read off where the line meets th compass

1 Let's find the direction from Point A to Point B. First, we turn the map so that its north arrow aligns with the compass's north arrow.

2 Now we put the compass over point A. We can see that Point B is southeast of Point A. This means we could get from A to B by using the compass to guide us southeast.

Using a compass to navigate

Let's practise using compass bearings by navigating our way round this map of the Android Islands in Cyberland.

1 The motor boat could get to the cafe via this course: three squares north, then four squares east. We write this as 3N, 4E.

2 The canoeist can reach the cave by following this course: 2E, 2S, 1W.

3 One way for the yacht to get to the harbour would be to sail 6N, 3W, 1N, 1W.

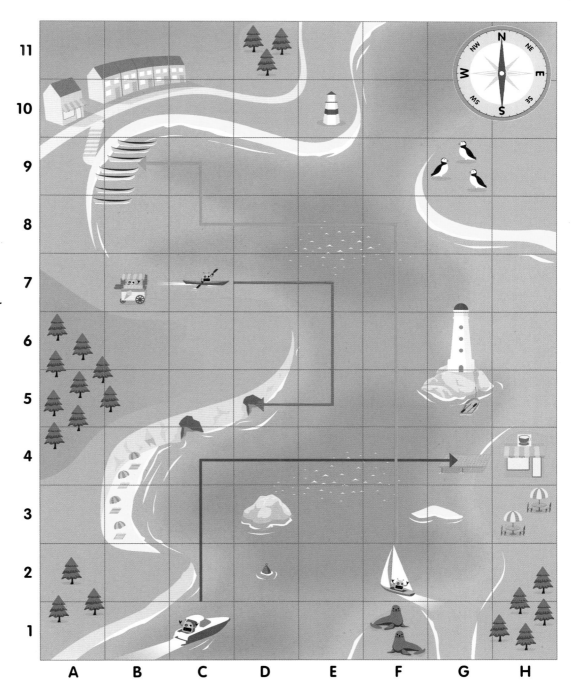

TRY IT OUT

Get your bearings

Now it's your turn to navigate your way around the Android Islands. Can you write directions for these journeys?

1 The lighthouse keeper wants an ice cream. Can you give directions to steer his boat to the ice cream cart?

2 Can you direct the motor boat to the puffins on Puffin Island?

3 If the yacht sailed a course of 1W, 2N, 2W, 1S, 1W, where would it reach?

4 If the canoeist paddled 3E, 6S, where would he end up?

Answers on page 320

Reflective symmetry

A shape has reflective symmetry if you can draw a line through it, dividing it into two identical halves that would fit exactly onto each other.

A line of symmetry is also called an axis of symmetry or mirror line.

How many lines of symmetry?

A symmetrical shape can have one, two, or lots of lines of symmetry. A circle has an unlimited number!

1 One vertical line of symmetry
This butterfly shape has only one line of symmetry. The shape is exactly the same on each side of the line. If you drew a line anywhere else on the shape, the two sides wouldn't be the same.

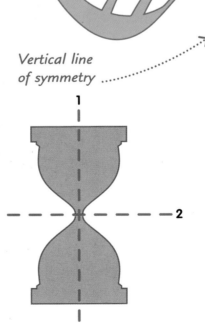

Each line of symmetry is numbered

Vertical line of symmetry

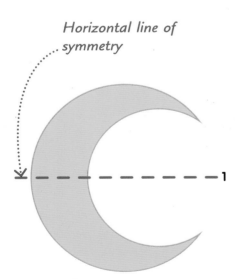

Horizontal line of symmetry

2 Horizontal line of symmetry
On this shape, the top and bottom halves are mirror images of each other.

3 Two lines of symmetry
This shape has both a horizontal and a vertical line of symmetry.

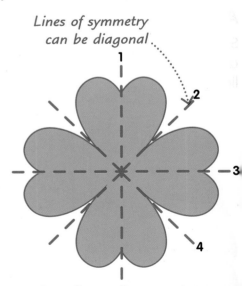

Lines of symmetry can be diagonal

4 Four lines of symmetry
This clover shape has one vertical, one horizontal, and two diagonal lines of symmetry.

Lines of symmetry in 2D shapes

Here are the lines of symmetry in some common 2D shapes.

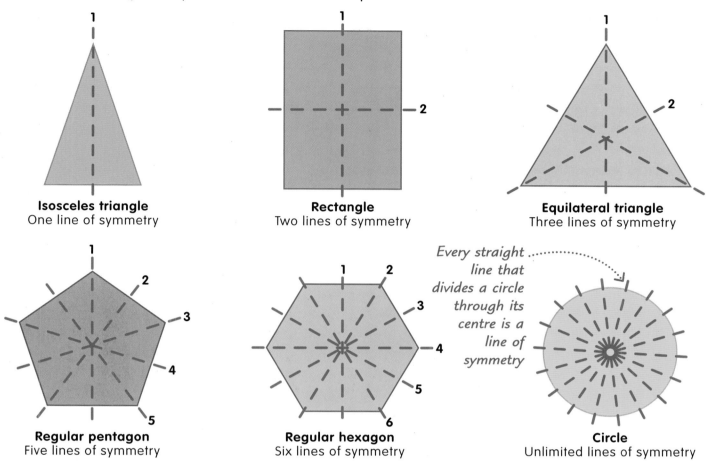

Isosceles triangle
One line of symmetry

Rectangle
Two lines of symmetry

Equilateral triangle
Three lines of symmetry

Regular pentagon
Five lines of symmetry

Regular hexagon
Six lines of symmetry

Every straight line that divides a circle through its centre is a line of symmetry

Circle
Unlimited lines of symmetry

Asymmetry

Some shapes are asymmetrical, which means they don't have any lines of symmetry. You can't draw a line anywhere on them to make a mirror image.

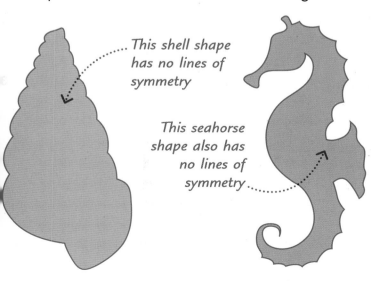

This shell shape has no lines of symmetry

This seahorse shape also has no lines of symmetry

Number symmetry

Look at each of these numbers. How many lines of symmetry does each one have? The answer will either be one, two, or none.

3 6 7 8

Answers on page 320

Rotational symmetry

We say that an object or shape has rotational symmetry if it can be turned, or rotated, about a point until it fits exactly into its original outline.

Centre of rotational symmetry

The point about which an object is rotated is called its centre of rotational symmetry.

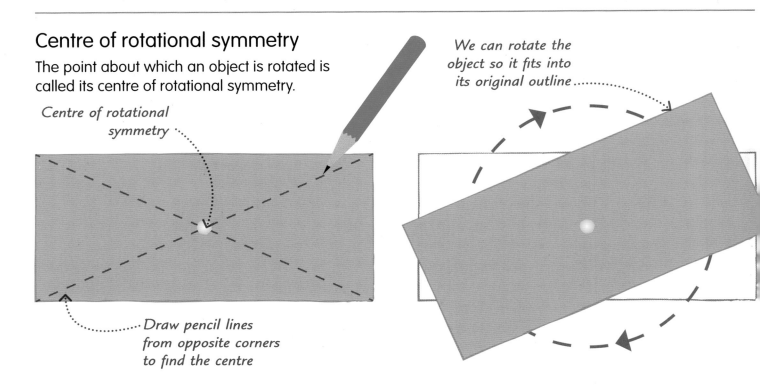

Centre of rotational symmetry

We can rotate the object so it fits into its original outline

Draw pencil lines from opposite corners to find the centre

1 Let's take a rectangular piece of card and put a pin through its centre, which is the point where the rectangle's two diagonals meet. Now let's draw around the outline of the rectangle.

2 If we rotate the rectangle around the pin, after half a turn it will fit exactly over the outline we drew. This means it has rotational symmetry. Another half turn will bring the rectangle back to its starting position.

TRY IT OUT

Symmetrical or not?

Three of these flower shapes have rotational symmetry. Can you spot the one that doesn't?

1   **2**   **3**   **4**

Answer on page 320

Order of rotational symmetry

The number of times a shape can fit into its outline during a full turn is called its order of rotational symmetry.

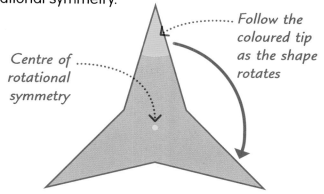

Follow the coloured tip as the shape rotates

Centre of rotational symmetry

1 Let's see how many times this three-pointed shape can fit into its outline. First, we rotate it until the yellow tip reaches the next point.

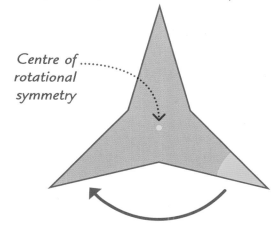

Centre of rotational symmetry

2 Now we rotate the shape again so the yellow tip moves to the next point.

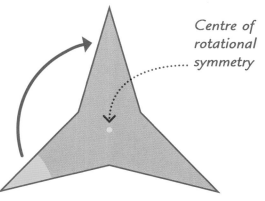

Centre of rotational symmetry

3 One more rotation and the yellow tip is back where it started. This shape can fit onto itself three times, so it has an order of rotational symmetry of 3.

Order of rotational symmetry in 2D shapes

Here are the orders of rotational symmetry for some common 2D shapes.

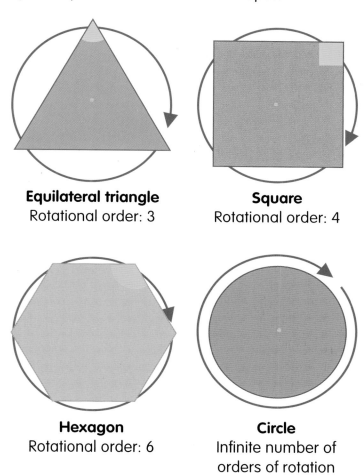

Equilateral triangle
Rotational order: 3

Square
Rotational order: 4

Hexagon
Rotational order: 6

Circle
Infinite number of orders of rotation

REAL WORLD MATHS

Symmetrical decoration

We often use rotational symmetry to make decorative patterns. In Islamic art, reflective and rotational symmetry are used to create intricate patterns on tiles for mosques and other buildings.

Reflection

In maths, we call a change in the size or position of an object a transformation. Reflection is a kind of transformation in which we make a mirror image of an object.

Reflection means flipping an object or shape over an imaginary line.

What is reflection?

A reflection shows an object or shape flipped so it becomes its mirror image across a line of reflection.

1 The original object is called the pre-image.

2 A reflection takes place over a line of reflection, like this one. It's also called the axis of reflection or mirror line.

3 The reflected version of the original shape or object is called the image.

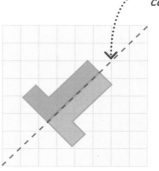

Lines of reflection

A shape and its reflected image are always on opposite sides of the line of reflection. Every point on the image is the same distance from the line of reflection as the pre-image. The line of reflection can be horizontal, vertical, or diagonal.

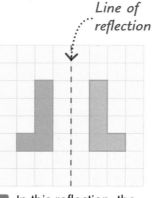

Line of reflection

1 In this reflection, the long sides of the image and pre-image are parallel to the line of reflection.

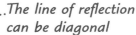

The line of reflection can be diagonal

2 This reflection is across a diagonal line. The sides of the shapes sit along the line of reflection.

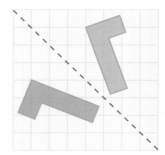

3 In this reflection, no part of the shape is parallel to or touching the line of reflection.

Drawing reflections

It's easier to draw reflections using grid or dot paper, which will help you to place the reflection accurately.

Vertical line of reflection

The line between the two points crosses the line of reflection at a right angle

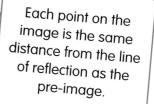

Each point on the image is the same distance from the line of reflection as the pre-image.

1 Let's try reflecting a triangle. First, draw a triangle on grid or dot paper. Label the vertices A, B, and C. Now draw a vertical line of reflection.

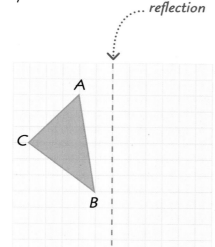

2 Count the squares from A to the line of reflection. Now count the same number of squares on the other side of the line and mark the point A'.

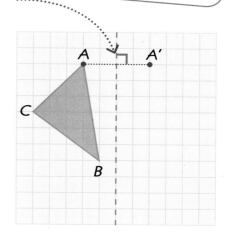

3 Do the same for the other two vertices of the triangle, marking the new points B' and C'.

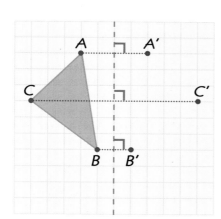

4 Finally, draw lines to join points A', B', and C'. You now have a new triangle that is a reflection of triangle ABC.

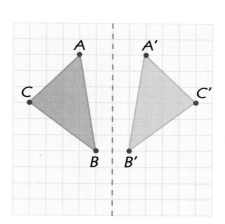

TRY IT OUT

Make a reflection pattern

You can use reflection to make symmetrical patterns. Draw a horizontal and a vertical line on grid paper to make four quadrants, and copy this design into the first quadrant. Then reflect it horizontally and vertically into each quadrant to complete the pattern.

Answer on page 320

Rotation

Rotation is a kind of transformation, in which an object or shape turns around a point called the centre of rotation. The amount we rotate the shape is called the angle of rotation.

Centre of rotation

The centre of rotation is a fixed point, which means it doesn't move. Let's look at what happens when we rotate the same shape clockwise around different centres of rotation.

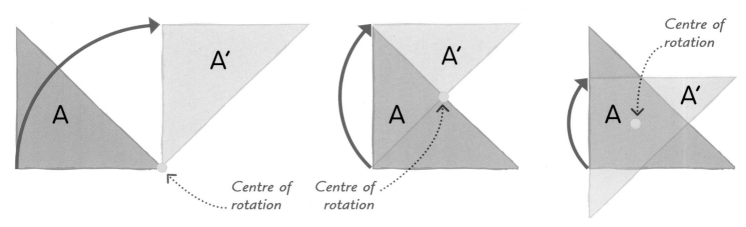

A   A'

Centre of rotation

A   A'

Centre of rotation

Centre of rotation

A   A'

1 First, let's rotate triangle A around one of its vertices to make a new triangle. We call the new triangle A'.

2 When we rotate A around the centre of its longest side, half of the new triangle overlaps the old one.

3 When A rotates around its centre, a different part of the new triangle overlaps the middle of the old one.

Angle of rotation

The angle of rotation is the distance that something rotates around a point, measured in degrees. Let's see what happens to this windmill sail when we rotate it by different amounts.

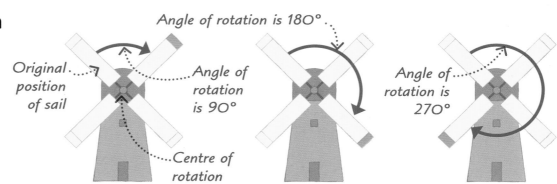

Angle of rotation is 180°

Original position of sail

Angle of rotation is 90°

Centre of rotation

Angle of rotation is 270°

1 This windmill sail has rotated through 90° (a right angle).

2 This time, the sail has rotated 180°, or two right angles.

3 Now the sail has rotated through 270°, or three right angles in total.

Rotation patterns

We can make patterns by rotating a shape lots of times around the same centre of rotation. This T shape makes different patterns depending on the centre and angle of rotation we choose.

Starting shape for all three patterns

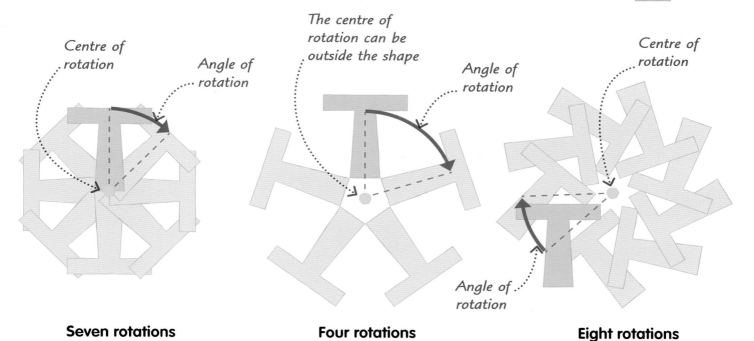

Centre of rotation

Angle of rotation

The centre of rotation can be outside the shape

Angle of rotation

Centre of rotation

Angle of rotation

Seven rotations
45° Angle of rotation

Four rotations
72° Angle of rotation

Eight rotations
40° Angle of rotation

TRY IT OUT

Make a rotation creation

All you need to make your own rotation pattern is some card and paper, a pin, a pair of scissors, and a pencil.

1 Draw a simple shape onto card and cut it out.

2 Put a pin through the the shape to make the centre of rotation.

3 Pin the shape to some paper and draw round the outline.

4 Rotate the shape a little and draw round it again. Repeat until you have a pattern you like!

Translation

A translation moves an object or shape into a new position by sliding it up, down, or sideways. Translation doesn't change its shape or size.

Translation is another kind of transformation, along with reflection and rotation.

What is translation?

Translation, like reflection or rotation, is a kind of transformation. With translation, the object and its image still look the same, because the original is not reflected, rotated, or re-sized – it just slides into a new position.

The original object or shape ..... is called the pre-image

The translated object or shape is called the image.....

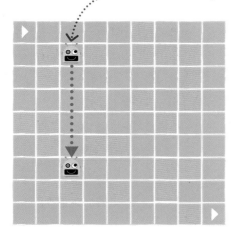

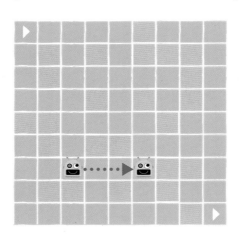

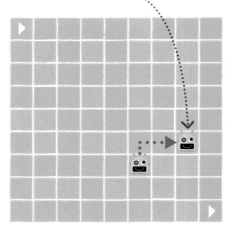

1 Look at the robot in this maze. It has moved vertically down by five squares.

2 This time, the robot has moved three squares horizontally to the right.

3 In this translation, the robot has moved one square up and two squares to the right.

REAL WORLD MATHS

Translation for tessellation

Translation is often used to make patterns called tessellations, which are identical shapes arranged together without leaving any gaps. This tessellation has been made by translating purple and orange cat shapes diagonally so that they interlock.

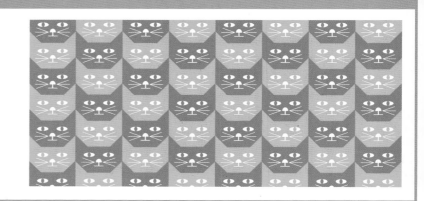

Using a grid to translate a shape

When we use a grid to translate a shape, we use the word "units" to describe the number of squares the shape is translated by. Let's translate a triangle!

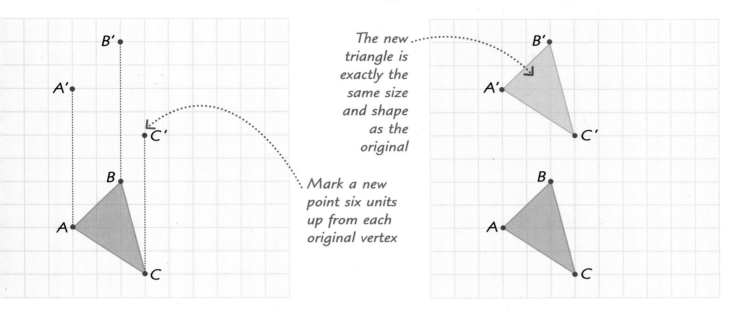

The new triangle is exactly the same size and shape as the original

Mark a new point six units up from each original vertex

1 Let's move the triangle up by six units. First, we label the vertices A, B, and C. Then we count up six units from each vertex and label the new points A', B', and C'.

2 Now use a ruler and pencil to join up the points you made to draw the new triangle A'B'C'.

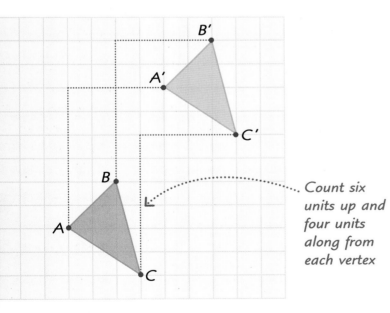

Count six units up and four units along from each vertex

3 To make a diagonal translation, count six units up, then four units to the right from each vertex. Plot the three new points and draw the new triangle A'B'C'.

TRY IT OUT

Triangle translations

How many different translations of the triangle are possible on this geoboard? We've shown one to get to you started – now it's over to you!

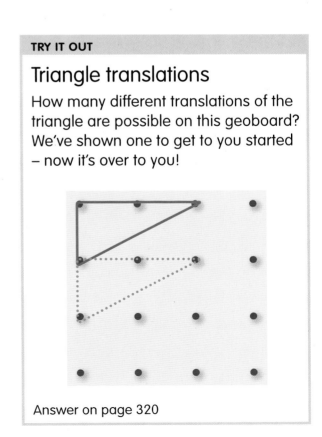

Answer on page 320

STATISTICS

Statistics is about collecting data and finding
out what it can tell us. The clearest way to
organise and analyse a large amount of data
is often to present it in a visual way, for
example by drawing a graph or chart.
We also use statistics to work out the chance,
or probability, that something will happen.

Data handling

Statistics is often called data handling. "Data" just means information. Statistics involves collecting, organizing, and presenting (displaying) data. It also involves interpreting the data – trying to understand what it can tell us.

1 We can collect data by carrying out a survey. In a survey, we ask a group of people questions and record their answers. These two survey robots are asking a class of school children which fruit they prefer.

2 Survey questions are often written on a form called a questionnaire. This is the robots' questionnaire. It asks children to choose between five fruits.

3 If there are several possible answers to a question, these may be listed on the questionnaire. There will be a tick box beside each answer so that it is quick and easy to record a response.

4 The answers the children give, before the data is organized, are called raw data.

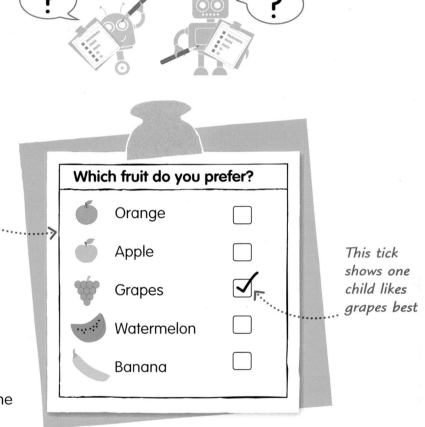

This tick shows one child likes grapes best

Voting

Another way to collect data is to hold a vote about something. You ask a question, and people give their answers – for example, by raising a hand. Then you count the number of raised hands. These robots are voting on whether they prefer nuts or bolts.

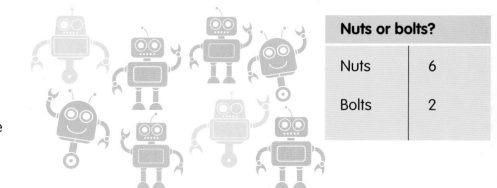

Nuts or bolts?	
Nuts	6
Bolts	2

What do we do with data?

Once data has been collected it needs to be organized and presented. Tables, charts, and graphs are quick ways of making data easy to read and understand.

Most popular fruit	
Type of fruit	**Number of children**
Orange	3
Apple	6
Grapes	8
Watermelon	2
Banana	5

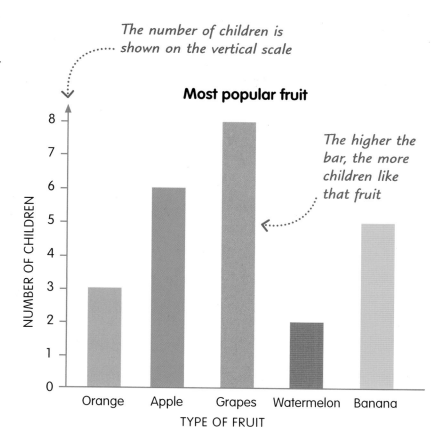

The number of children is shown on the vertical scale

Most popular fruit

The higher the bar, the more children like that fruit

1 This table, called a frequency table, shows the number of children that preferred each type of fruit.

2 A bar chart, also called a bar graph, is a diagram that shows data without the need for lots of words or lists of numbers.

Data sets

A set is a collection of data. It can be a group of numbers, words, people, events, or things. Sets can be divided into smaller groups called subsets.

1 The class of children that the robots asked about their favourite fruit is a set. The class contains 24 children, a mixture of boys and girls.

2 The eight boys (shown in red) are a subset of the class. The 16 girls (green) are also a subset. Together, they form the set of the whole class.

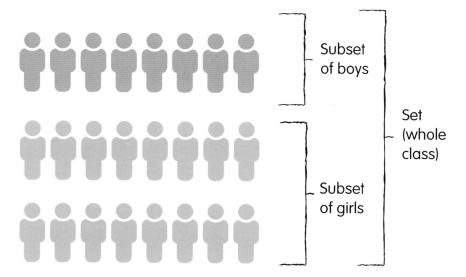

Subset of boys

Set (whole class)

Subset of girls

Tally marks

We can use tally marks to count things quickly when we're collecting data, such as answers to a survey question. A tally mark is a vertical line that represents one thing counted.

You make a tally mark for each thing that you count.

1 Draw a tally mark to show each result you record. For every fifth tally mark, draw a line across the previous four. This is how the numbers one to five look when written as tally marks.

| | || | ||| | |||| | |||| |
| 1 | 2 | 3 | 4 | 5 |

2 Arranging tally marks into groups of five helps you to work out the total quickly. First, count all the groups of five, then add any remaining tallies. This is how 18 looks in tally marks.

$$5 + 5 + 5 + 3 = 18$$

3 A tally chart, such as the one below, uses tally marks to show the results of a survey.

Each tally mark represents one child

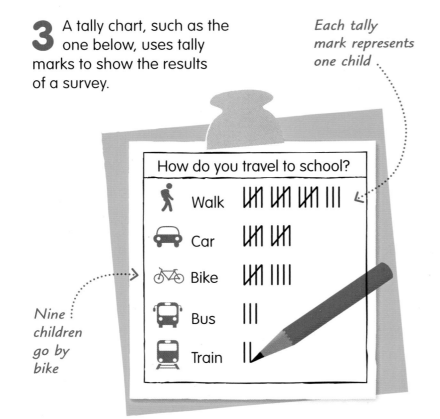

How do you travel to school?

Walk																
Car									
Bike									
Bus				
Train			

Nine children go by bike

REAL WORLD MATHS

Other tally marks

Tally marks vary across the world. In some Asian countries, they are based on a Chinese symbol made up of five strokes.

| ─ | ⊤ | 丆 | 讠匚 | 讠匸 |
| 1 | 2 | 3 | 4 | 5 |

In parts of South America, four lines are drawn to make a square, then a diagonal line is drawn across it for the fifth mark.

| | | ⌐ | ⊓ | □ | ◩ |
| 1 | 2 | 3 | 4 | 5 |

Frequency tables

A frequency table is a way of summarizing a set of data. The table shows you exactly how many times each number, event, or item occurs in the set of data.

The frequency of something tells you how often it happens.

1 You can create a frequency table by counting the tally marks in a tally chart and writing the totals in a separate column.

2 This frequency table is based on the survey of how children travelled to school. The frequency column shows you how many children used each type of transport.

How we travelled to school

Transport	Tally	Frequency
Walk	ЖЖ ЖЖ ЖЖ III	18
Car	ЖЖ ЖЖ	10
Bike	ЖЖ IIII	9
Bus	III	3
Train	II	2

Count the tally marks and put the totals in this column

3 Frequency tables don't always look the same. The table here uses the same data as the one above, but it doesn't include the tally marks. This makes the table simpler and easier to understand.

Travelling to school

Transport	Frequency
Walk	18
Car	10
Bike	9
Bus	3
Train	2

Frequency is shown only as numbers

The museum is closed on Mondays

4 Some frequency tables split up data so it reveals more information. This table tells you how many adults and children visited a dinosaur museum each day during one week. It also tells you the total number of visitors there were (adults + children) each day.

Day	Adults	Children	Total
Monday	0	0	0
Tuesday	301	326	627
Wednesday	146	348	494
Thursday	312	253	565
Friday	458	374	832
Saturday	576	698	1274
Sunday	741	639	1380

Dinosaur Museum visitor numbers

Carroll diagrams

A Carroll diagram shows how a set of data, such as a group of people or numbers, has been sorted. Carroll diagrams sort data using conditions called criteria (the singular is criterion).

Carroll diagrams sort data into boxes.

1 A criterion is like a yes/no question. Let's use a simple criterion to sort the group of 12 animals below. This is the criterion we'll use: "is the animal a bird or not?"

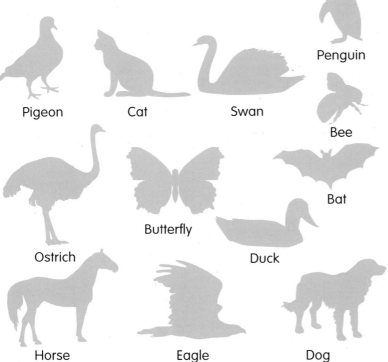

Penguin

Pigeon   Cat   Swan

Bee

Butterfly

Bat

Ostrich

Duck

Horse   Eagle   Dog

2 This Carroll diagram uses our bird/not a bird criterion to sort the animals into two boxes. We put all the birds into the box on the left. Those animals that aren't birds go in the box on the right.

Bird	Not a bird
Pigeon	Butterfly
Duck	Cat
Penguin	Bat
Eagle	Bee
Swan	Dog
Ostrich	Horse

All the animals fit into one box or the other

3 To further sort our group of animals using the Carroll diagram, we can add a new criterion: "is it a flying animal or not?" To fit into any box, an animal must now meet two criteria.

Animals that are birds and can fly

Animals that are not birds but can fly

	Bird	Not a bird
Flying	Pigeon Eagle Swan Duck	Butterfly Bat Bee
Not flying	Penguin Ostrich	Dog Horse Cat

Animals that are birds but can't fly

Animals that are not birds and cannot fly

Sorting numbers

Carroll diagrams can sort numbers and show relationships between them. This diagram sorts the set of the numbers from 1 to 20 into even, odd, prime, and not prime numbers.

Subset of prime numbers from 1 to 20

Subset of numbers from 1 to 20 that are not prime numbers

1 If we read down the first column (yellow), we see all the prime numbers. The second column (green) shows all the non-primes.

	Prime number	Not a prime number
Even number	2	4 6 8 10 12 14 16 18 20
Odd number	3 5 7 11 13 17 19	1 9 15

Subset of even numbers from 1 to 20

Subset of odd numbers from 1 to 20

2 When we read across the first row (blue) we see all the even numbers. The second row (red) lists the odd numbers.

	Prime number	Not a prime number
Even number	2	4 6 8 10 12 14 16 18 20
Odd number	3 5 7 11 13 17 19	1 9 15

Subset of even numbers from 1 to 20 that are not primes

3 All the even numbers that are not primes are in the box in the top right corner (orange). Odd numbers that are not primes are in the box beneath (pink).

	Prime number	Not a prime number
Even number	2	4 6 8 10 12 14 16 18 20
Odd number	3 5 7 11 13 17 19	1 9 15

Subset of odd numbers from 1 to 20 that are not primes

Subset of even prime numbers from 1 to 20

4 The only even prime number, 2, is shown in the box in the top left corner (yellow). The box beneath (green) shows all the odd prime numbers.

	Prime number	Not a prime number
Even number	2	4 6 8 10 12 14 16 18 20
Odd number	3 5 7 11 13 17 19	1 9 15

Subset of odd prime numbers from 1 to 20

Venn diagrams

A Venn diagram shows the relationships between different sets of data. It sorts the data into overlapping circles. The overlaps show what the sets have in common.

Venn diagrams show sets of data as overlapping circles.

1 Remember, a set is a collection of things or numbers, or a group of people. For example, a set might be the foods you like or the dates of your family's birthdays. This group of eight friends forms a set. Most of them do activities after school.

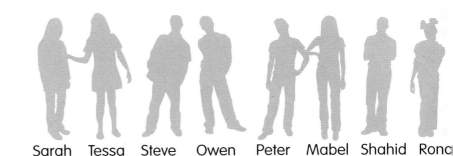

Sarah   Tessa   Steve   Owen   Peter   Mabel   Shahid   Rona

2 Each thing or person in the set is called a member or element of the set. Sets are often shown with a circle drawn around them. Here is the set of friends.

Shahid   Tessa

Sarah   Rona

Owen   Mabel

Peter   Steve

.... *Set of friends*

.... *Each friend is a member of the set*

3 There are three after-school activities that the friends do: music lessons, art classes, and football practice. We can put the friends into smaller sets, according to which after-school activities they do.

Friend who does no after-school activities

Shahid
Sarah
Rona
Steve

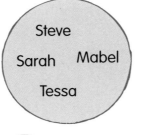

Steve
Sarah   Mabel
Tessa

Steve
Rona   Tessa
Peter

Owen

 **Music lesson**   **Art class**    **Football practice**   **No activities**

STATISTICS • **VENN DIAGRAMS**

4 Let's join the music and soccer sets together so that their circles overlap. When we join two sets, it's called a union of sets. We've now made a Venn diagram.

5 An overlap between two sets is an intersection. It shows when something belongs to more than one set. This intersection shows that Rona and Steve do both activities.

6 Now let's join the art set to the other two sets, so that all three sets overlap. If we look at the intersections, we can see which friends do more than one activity.

7 Our three-set Venn diagram includes only seven of the eight friends. Owen doesn't do any after-school activities, so he doesn't belong to any of those sets.

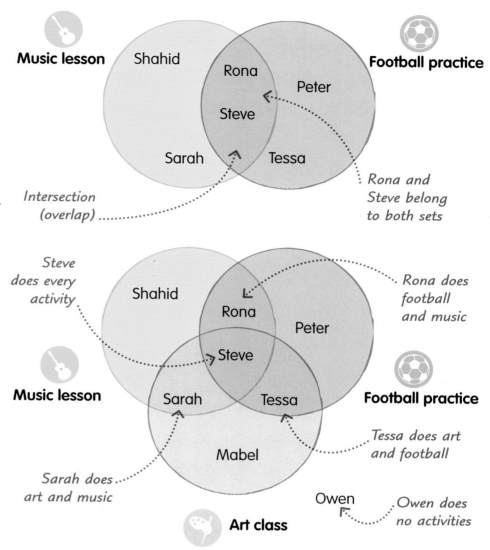

Music lesson **Football practice**

Shahid, Rona, Peter, Steve, Sarah, Tessa

Intersection (overlap)

Rona and Steve belong to both sets

Steve does every activity

Rona does football and music

Music lesson **Football practice**

Shahid, Rona, Peter, Steve, Sarah, Tessa, Mabel

Sarah does art and music

Tessa does art and football

Owen *Owen does no activities*

Art class

The universal set

The universal set is the set that contains everybody or everything that is being sorted, including those not in the overlapping sets.

1 To show the universal set, we draw a box around all the intersecting circles in our diagram.

2 The box must include Owen. Even though he is not in any of the after-school activity sets, he is still part of the group being sorted.

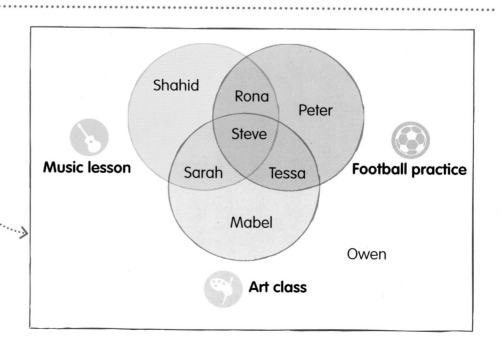

Shahid, Rona, Peter, Steve, Sarah, Tessa

Music lesson **Football practice**

Mabel

Owen

Art class

Averages

An average is a kind of "middle" value used to represent a set of data. Averages help you to compare different sets of data, and to make sense of individual values within a data set.

The average is the value that's most typical of a set of data.

1 The average age of the Reds football team is 10. Not all the players are 10 years old – some are 9 and some are 11. But 10 is the age that is typical of the team as a whole.

Average age = 10

Player's a.

2 The average age of the Blues football team is 12. Comparing the two averages, we can see that the Blues team is, typically, older than the Reds.

Average age = 12

3 An average can also tell us if an individual value is typical of the data set or unusual. For example, the Reds' average age of 10 can tell us if these three players aged 9, 10, and 11 are typical of the team or not.

age 9 — Not typical of team

age 10 — Typical of team

age 11 — Not typical of team

Types of average

We can use three different types of average to describe a set of data, such as the heights of a group of giraffes. They are called the mean, the median, and the mode. Each one tells us something different about the group. But they all use a single value to represent the group as a whole. To find out more, see pages 277-79.

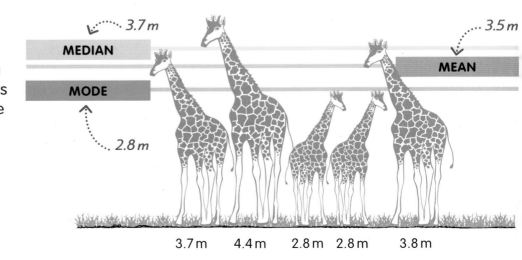

MEDIAN 3.7 m
MODE 2.8 m
MEAN 3.5 m

3.7 m   4.4 m   2.8 m   2.8 m   3.8 m

The mean

When people talk about the average, they are usually talking about the mean. We work it out by adding up the individual values in a group and dividing the total by the number of values.

The mean is the sum of all values divided by the number of values.

1 Let's find the mean height of this group of five giraffes.

2 First, we add up all the heights of the giraffes:
3.7 + 4.4 + 2.8 + 2.8 + 3.8 = 17.5

3 Now divide the total height by the number of giraffes: 17.5 ÷ 5 = 3.5

4 So, the mean height of these giraffes is 3.5 m.

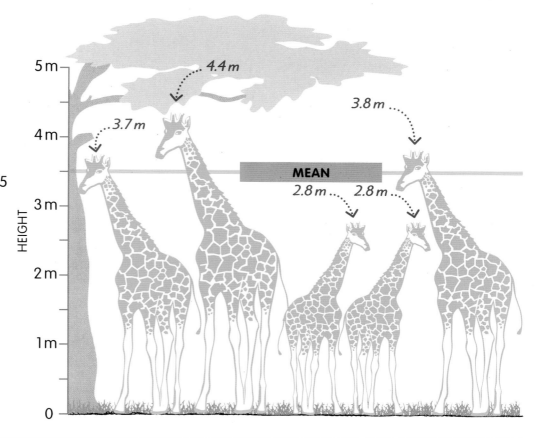

HEIGHT

5 m — 4.4 m

4 m — 3.7 m   3.8 m

          MEAN

3 m —   2.8 m   2.8 m

2 m —

1 m —

0

TRY IT OUT

Is it hot today?
Or just average?

Weather forecasts often mention average or mean temperatures. Here are the midday temperatures for a week. Let's work out the mean temperature.

Answer on page 320

1 First, add up all the individual temperatures.

2 Then count the number of temperatures.

3 To find the mean, divide the total of the temperatures by the number of temperatures.

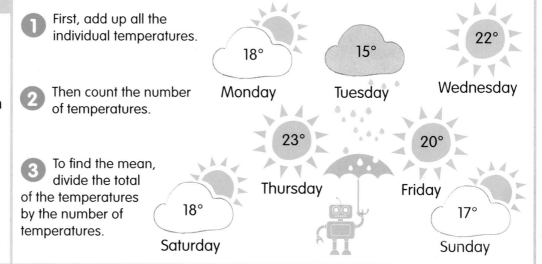

18° Monday

15° Tuesday

22° Wednesday

23° Thursday

20° Friday

18° Saturday

17° Sunday

The median

The median is simply the middle value in a set of data when all the values are arranged in order, from smallest to largest or from largest to smallest.

The median is the middle value when all the values are arranged in order.

1 Take another look at our group of giraffes. This time, let's work out the median height.

2 Write down the heights in order, starting with the shortest: 2.8, 2.8, 3.7, 3.8, 4.4

3 Now find the middle height. This is 3.7, because there are two heights that are shorter and two that are taller.

4 So, the median height is 3.7 m.

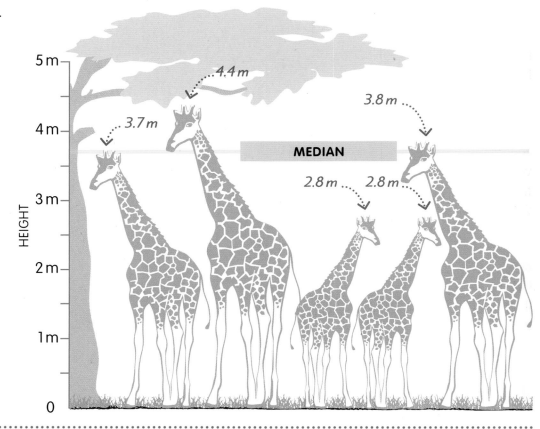

4.4 m

3.7 m

3.8 m

MEDIAN

2.8 m   2.8 m

HEIGHT

5 m
4 m
3 m
2 m
1 m
0

Add one giraffe

What happens if another giraffe, with a height of 4.2 m, joins the group to make six giraffes? With an even number of giraffes, there's no one middle height. We can still find the median by working out the mean of the middle two heights.

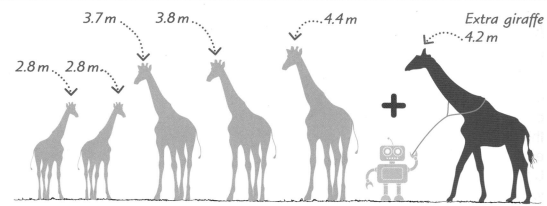

2.8 m   2.8 m   3.7 m   3.8 m   4.4 m   Extra giraffe 4.2 m   +

1 First, let's arrange the heights of our six giraffes in order: 2.8, 2.8, 3.7, 3.8, 4.2, 4.4

2 The two middle heights are 3.7 and 3.8. Now let's work out their mean: (3.7 + 3.8) ÷ 2 = 3.75

3 Adding one more giraffe has changed the median height to 3.75 m.

The mode

The mode is the value that occurs most often in a set of data. It is also called the modal value. Sometimes a set of data has more than one mode.

To find the mode, look for the value that occurs most often. It often helps to arrange the values in order.

1 We've worked out the mean and median heights of the giraffes. Now let's find the mode.

2 It's easier to see the most frequent value if we put the heights in order, from shortest to tallest: 2.8, 2.8, 3.7, 3.8, 4.4

3 Then we look at the list of heights to find the height that occurs most often. This is 2.8, which occurs twice.

4 So, the mode of the heights is 2.8 m.

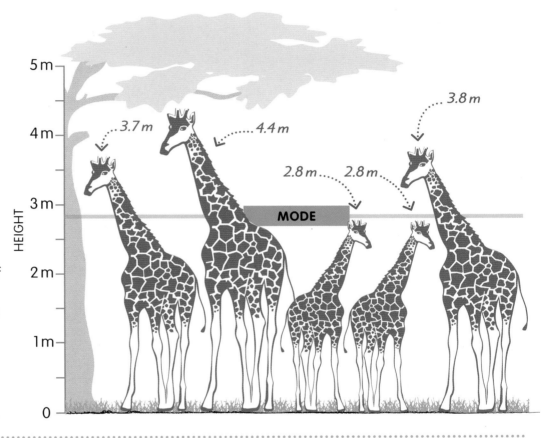

Multiple modes

When there are two or more values that are equally common and occur more often than the other values, then each of them is a mode. Let's see what happens when we add an extra giraffe, with a height of 4.4 m, to our group.

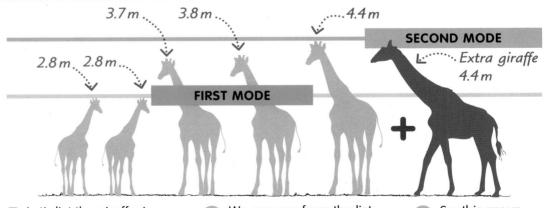

1 Let's list the giraffes' heights in order again, from shortest to tallest: 2.8, 2.8, 3.7, 3.8, 4.4, 4.4

2 We can see from the list that 2.8 and 4.4 both occur twice, while the other heights occur only once.

3 So, this group of heights now has two modes: 2.8 m and 4.4 m.

The range

The spread of values in a set of data is known as the range. It's the difference between the smallest and largest values in the set. Like averages, the range can be used to compare sets of data.

1 Let's find the range of our giraffes' heights. First, we'll write down their heights in order, from shortest to tallest. This gives us: 2.8, 2.8, 3.7, 3.8, 4.4

2 Now let's find the shortest and tallest heights. These are 2.8 m and 4.4 m.

3 Next, subtract the shortest height from the tallest. This gives 4.4 − 2.8 = 1.6

4 So, the range of the giraffes' heights is 1.6 m.

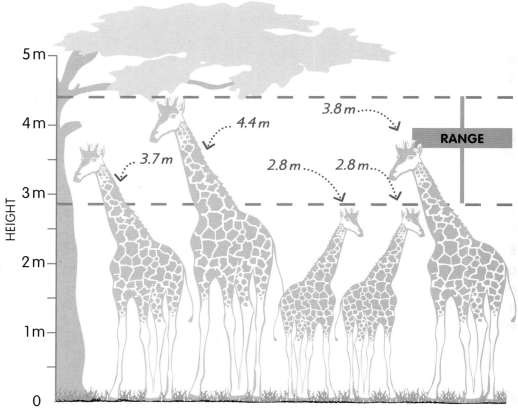

TRY IT OUT

Roll the dice, find the average

Don't worry if you haven't got a group of giraffes handy to help you understand averages; you can use dice instead. For these investigations, all you need is two dice.

1 Roll both dice. Write down the total number of spots.

2 Do this 10 times.

3 Calculate the mean and find the mode, median, and range for the dice rolls.

4 What if you roll the dice 20 times? Do you get the same mean, mode, median, and range?

To find the range, subtract the smallest value from the largest. The result is the range.

Using averages

Whether it's best to use the mean, median, or mode depends on the values in your data and the type of data involved. The range is helpful if the mean, median, and mode are all the same.

Avoid the mean if one value is a lot higher or lower than the others.

1 Use the mean if the values in a set of data are fairly evenly spread. Here, you can see the savings of five children. The mean (total savings ÷ number of children) is £66.00 ÷ 5 = £13.20

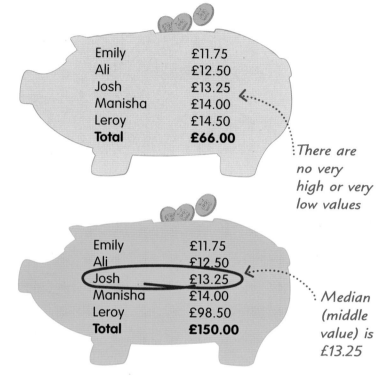

Emily	£11.75
Ali	£12.50
Josh	£13.25
Manisha	£14.00
Leroy	£14.50
Total	**£66.00**

There are no very high or very low values

2 The mean can be misleading if one value is much higher or lower than the rest.

3 For example, let's see what happens if Leroy saves £98.50, not £14.50. Now the mean is: £150 ÷ 5 = £30.00, which makes it seem like the others are saving much more than they really are. In this case, it's better to use the median (middle value) of £13.25. This is much closer to the amount that most of the children save.

Emily	£11.75
Ali	£12.50
Josh	£13.25
Manisha	£14.00
Leroy	£98.50
Total	**£150.00**

Median (middle value) is £13.25

4 The mode (most common value) can be used with data that isn't numbers. For example, in a survey of the colours of cars spotted, the mode might be blue.

Blue cars were seen most often

Using the range

The range (the spread of values) can be useful for showing a difference between data sets when their mean, median, and mode are the same.

1 Two football teams each scored 20 goals in five games. The mean goals scored per match for both teams is 4 (20 ÷ 5 = 4).

2 The median (middle value) for each team is also 4 goals. So, too, is the mode (most common value), since both teams scored 4 goals twice.

3 The range is different. It's 8 – 1 = 7 goals for the Reds. For the Blues, it's 6 – 1 = 5 goals. So, the Reds' data has a wider spread of values.

Goals scored	
Reds	Blues
8	6
4	5
4	4
3	4
1	1
Total: 20	Total: 20

Pictograms

In a pictogram, or pictograph, small pictures or symbols are used to represent data. To divide the data into groups, the pictures are usually placed in columns or rows.

In a pictogram, pictures stand for numbers.

Always give your pictogram a title ....

There are six symbols, so children saw six pigeons

Children saw more starlings than robins

Choose an appropriate symbol to represent your data

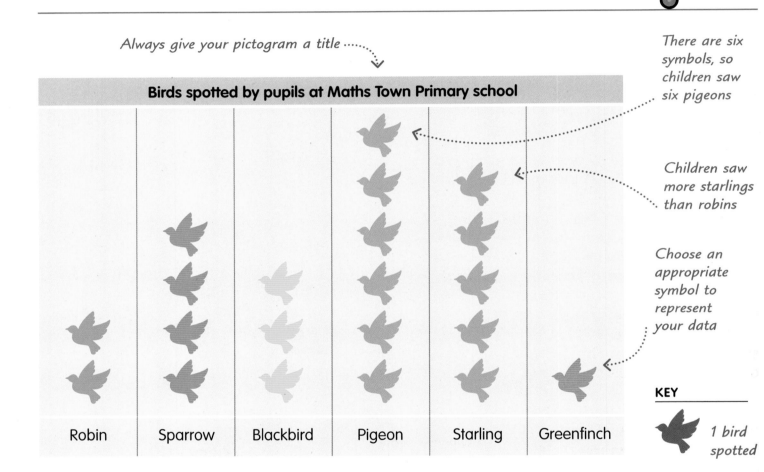

Birds spotted by pupils at Maths Town Primary school

| Robin | Sparrow | Blackbird | Pigeon | Starling | Greenfinch |

KEY

1 bird spotted

1 Let's look at this simple pictogram. It shows the results of a survey of the types and numbers of birds seen by children at a primary school.

2 The set of data shown in the pictogram is all the birds seen. Each type of bird is a subset of this larger set. For example, there is one subset for blackbirds.

3 A pictogram must have a key to explain what one symbol or picture stands for. Here, the key shows that one symbol means 1 bird spotted.

4 Count the symbols in a column to find out how many birds of that type the children saw. This is the frequency of the subset. For example, the frequency of blackbirds is three.

Using large numbers

When a pictogram needs to show large numbers, each picture or symbol can represent more than one. In this pictogram, each symbol stands for two people who visited a library. Half a symbol represents one person.

KEY

👤 = 2 people

16 people over 60 visited the library

Age	Number of people
Visitors to Maths Town Library	
Over 60 years	👤 👤 👤 👤 👤 👤 👤 👤
19–60 years	👤 👤 👤 👤 👤 👤 👤 👤 👤
11–18 years	👤 👤 👤 👤 ⦙
5–10 years	👤 👤 👤 👤 👤 👤
Under 5 years	👤 👤 👤 👤 👤 👤 ⦙

A half symbol represents one person

1 To find the number of visitors in a particular age group, count the full symbols in that row, multiply by two, and add one if there's a half symbol.

2 How many people aged 11 to 18 visited? There are four full symbols plus one half symbol. So the calculation is: (4 × 2) + 1 = 9

TRY IT OUT

Make a pictogram

Use this frequency chart to make a pictogram showing how much time Leroy spends playing video games during the school week.

1 Design a symbol or draw a picture to use on your pictogram. It must be suitable and easy to understand.

2 How many minutes will your symbol represent? Will you use half symbols as well as full ones?

3 Will you arrange your symbols in vertical columns or horizontal rows?

Leroy's gaming	
Day	**Gaming time**
Monday	30 minutes
Tuesday	60 minutes
Wednesday	15 minutes
Thursday	45 minutes
Friday	75 minutes

Block graphs

A block graph is a kind of graph in which one block, usually a square, is used to represent one member of a group or set of data. The blocks are stacked in columns.

Block graphs show data as stacks of square blocks.

1 This tally chart shows the results of a survey that asked children which fruit they liked best. Let's use the data to make a block graph.

2 Each tally mark shows that one child chose that fruit.

Which fruit do you prefer?

🍎	Orange	|||
🍎	Apple	卌 I
🍇	Grapes	卌 |||
🍈	Watermelon	||
🍌	Banana	卌

Tally marks record frequency of the data

Six children preferred apples

3 We draw a square block on our graph for every tally mark on the chart. All the blocks must be the same size.

4 We stack the blocks on top of each other in columns. Leave gaps between the columns. The number of blocks in a column shows how many times that fruit was chosen (the frequency).

Each block tells you that one child chose that fruit

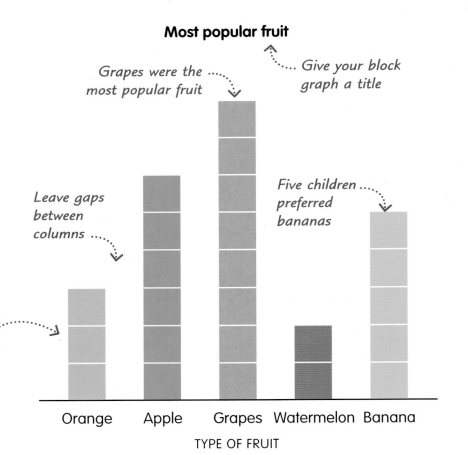

Most popular fruit

Grapes were the most popular fruit

Give your block graph a title

Leave gaps between columns

Five children preferred bananas

Orange　Apple　Grapes　Watermelon　Banana

TYPE OF FRUIT

Bar charts

A bar chart uses bars or columns to represent groups or sets of data. The size of each bar shows the frequency of the data. Bar charts are also called bar graphs and column graphs.

> The height or length of a bar shows the frequency.

1 Let's look at this bar chart. It uses data from a survey of car colours. The bars are all the same width, separated by gaps.

2 The chart is framed on two sides by lines called axes. The bars for car colours sit on the horizontal axis. A scale on the vertical axis shows the number of cars seen (the frequency).

3 To find out how many white cars were seen, look across from the top of the White bar to the vertical axis. Then read the number (5) off the scale.

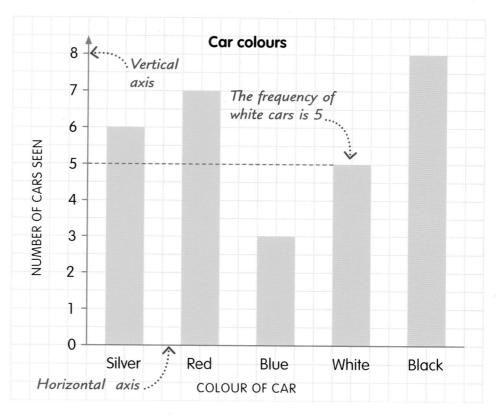

4 We can redraw the same chart so that the bars are horizontal, going across the chart, rather than vertical.

5 The car colours are now along the vertical axis, while the number of cars (the frequency) can be read off the horizontal axis.

The scale is now along the horizontal axis

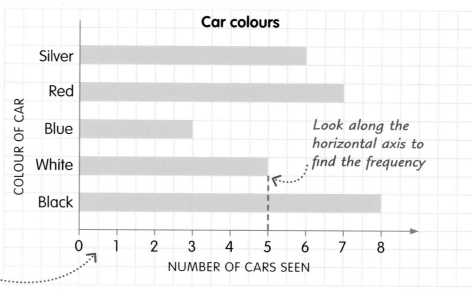

Drawing bar charts

To draw a bar chart you need a pencil, a ruler, an eraser, coloured pens, pencils, or crayons, and squared paper or graph paper. Most importantly, you need some data!

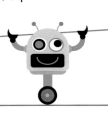

> Draw your bar charts on squared paper.

1 Let's use the data in this frequency table. It shows the results of a survey of instruments played by a group of children.

2 It's best to draw our bar chart on paper marked with small squares. This makes it easier to mark a scale and draw the bars.

3 First, we draw a horizontal line for the x axis and a vertical line for the y axis.

4 Next we draw marks on the x axis to show the width of the bars that represent the different instruments. All the bars must be of the same width. Let's make ours 2 small squares wide.

5 Now let's add a scale to the y axis to represent the number of children. We need a scale that covers the range of numbers on the table but doesn't make our chart look stretched or squashed. A scale from 0 to 8 works well here.

| What instrument do you play? ||
Instrument	**Children**
Guitar	7
Violin	6
Trumpet	3
Flute	4
Piano	5

Numbers in this column show the frequency

Squared paper

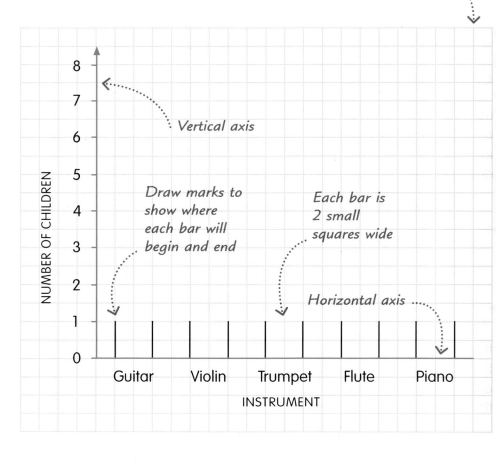

Vertical axis

Draw marks to show where each bar will begin and end

Each bar is 2 small squares wide

Horizontal axis

NUMBER OF CHILDREN

INSTRUMENT

6 Now let's start drawing the bars for our instruments. The first frequency in the table is 7, which represents the number of children who play guitar.

7 We find 7 on the vertical scale of the y axis. Next, we draw a short horizontal line level with 7. It must be exactly above the marks we made for the guitar bar on the x axis. We'll make the line 2 small squares long, the same as the distance between the markers.

8 Then we do the same for all the other instruments.

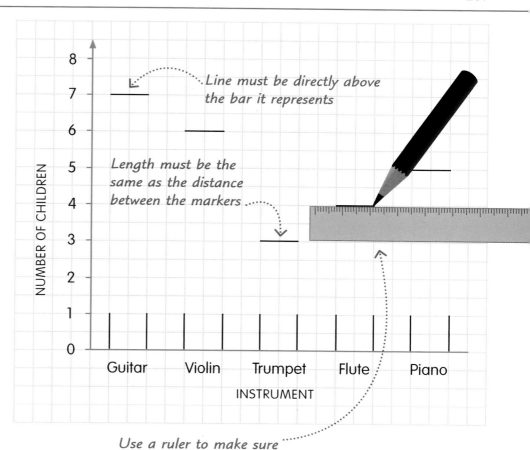

Line must be directly above the bar it represents

Length must be the same as the distance between the markers

Use a ruler to make sure your lines are straight

The two vertical lines meet the horizontal line to form a bar.

Give your bar chart a title

9 To complete the guitar bar, we draw two vertical lines up from its markers on the x axis. The lines join up with the ends of the horizontal line we drew earlier.

10 Then we do the same for all the other instruments.

11 Finally, let's colour in the bars. The bars can be all the same colour if you want. But if we make the bars different colours, it may make the chart easier to understand.

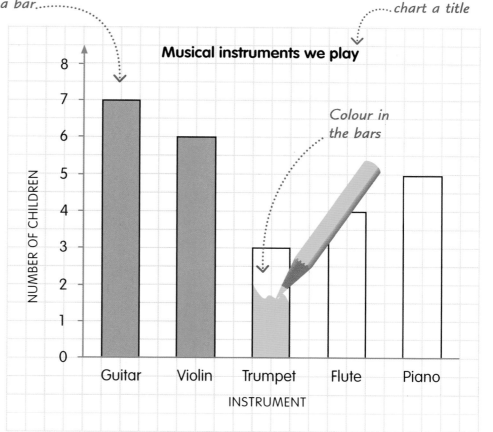

Musical instruments we play

Colour in the bars

Line graphs

On a line graph, frequencies or measurements are plotted as points. Each point is joined to its neighbours by straight lines. A line graph is a useful way to present data collected over time.

Line graphs are great for showing data over a length of time.

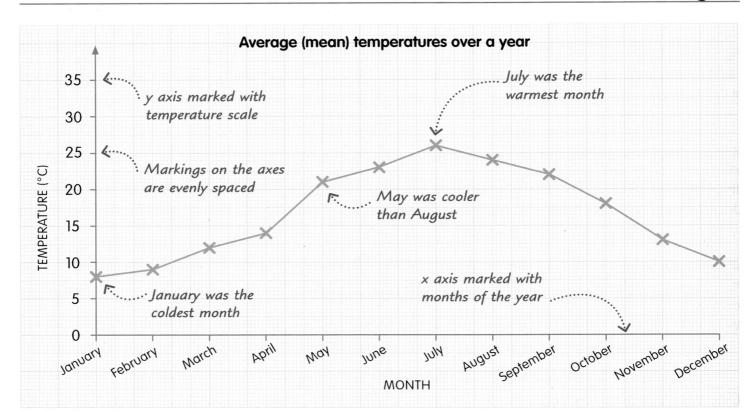

Average (mean) temperatures over a year

- 35 — *y axis marked with temperature scale*
- 25 — *Markings on the axes are evenly spaced*
- *January was the coldest month*
- *May was cooler than August*
- *July was the warmest month*
- *x axis marked with months of the year*

TEMPERATURE (°C)

MONTH

1 Let's look at this line graph. It shows the average monthly temperatures recorded in Maths Town over one year.

2 The months of the year are listed on the horizontal x axis, and a temperature scale runs up the vertical y axis.

3 The average temperature for each month is plotted with an "x". All the crosses are linked to form a continuous line.

4 The graph makes it easy to see which were the warmest and coldest months of the year. It also lets us compare the temperatures in different months.

REAL WORLD MATHS

Counting the beats

A heart monitor is a machine that records how fast your heart is beating. It shows the data as a line graph like a wiggly line on a screen or print out.

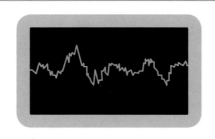

Reading line graphs

This graph tells us how Jacob grew between the ages of 2 and 12. We can see how tall he was at any age by going up from the x axis to the line, then across to the y axis. We can also estimate his height between yearly measurements.

1 Let's see how tall Jacob was aged 6. We find 6 on the x axis and then go straight up.

2 When we meet the green line, we go straight across to the y axis. This shows us that Jacob was 110 cm tall at age 6.

3 We can also work out Jacob's height at age 9½. Going up and across, the y axis tells us he was probably 132 cm tall.

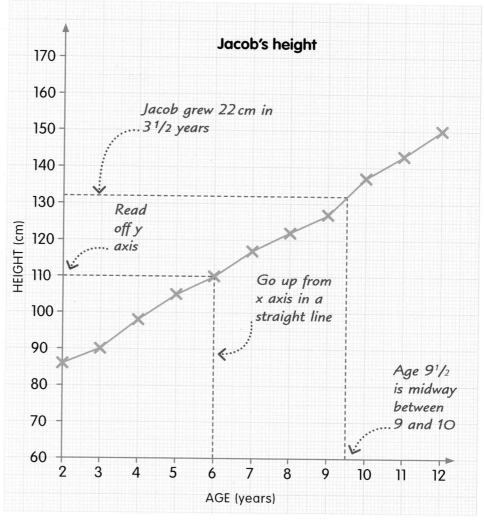

Conversion graphs

A conversion graph uses a straight line to show how two units of measurement are related.

1 This graph has kilometres on the x axis and miles on the y axis. The line lets us convert from one unit to the other.

2 To change 80 km into miles, we go along the x axis until we reach 80. Then we go up to the line and across to the y axis, where we read off 50 miles.

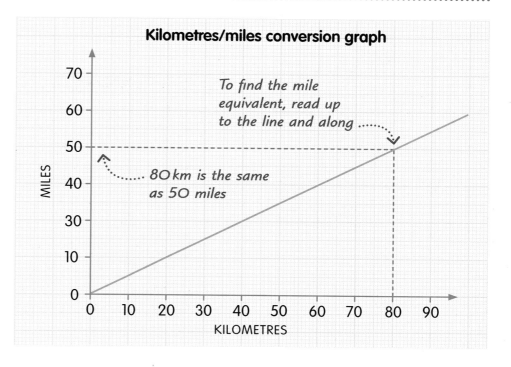

Drawing line graphs

A pencil, ruler, graph paper, and some data are all that's needed to draw a line graph. We plot data on the graph, usually as crosses. Then we join up the crosses to create a continuous line.

1 A class of school children recorded the outside temperature every hour as part of a science experiment. Let's use the data from this table to draw a line graph.

2 We'll use special graph paper marked with small squares. It will help us to plot data and draw lines accurately.

3 First, we need to draw our x and y axes. Time always goes along the horizontal x axis of a line graph. We mark and write the hours of the day along this axis, starting with 0800.

4 Temperature goes along the vertical y axis. We need to add a scale that covers the highest and lowest values in the table (the range). A scale from 0 to 18°C works well. Let's mark every two degrees, otherwise the scale will look too crowded.

5 We'll label the horizontal x axis "Time" and the vertical y axis "Temperature (°C)".

Hourly temperatures	
Time	**Temperature (°C)**
0800	6
0900	8
1000	9
1100	11
1200	12
1300	15
1400	16
1500	15
1600	13

The numbers in this column show the temperature at each hour

Graph paper

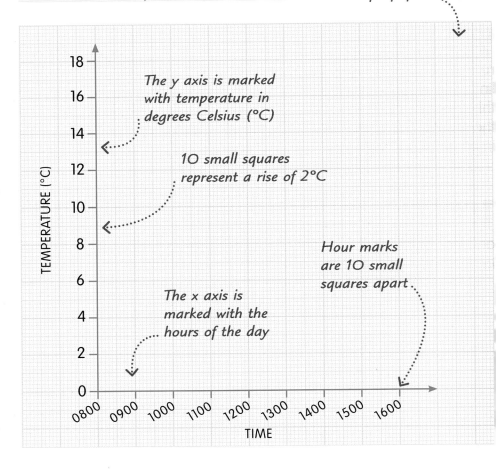

The y axis is marked with temperature in degrees Celsius (°C)

10 small squares represent a rise of 2°C

Hour marks are 10 small squares apart

The x axis is marked with the hours of the day

6 Now we can plot the data on our graph. Let's take each temperature in order and find its position on the graph.

7 The first temperature is 6°C at 0800. We go up the y axis from the 0800 marker on the x axis until we get to 6. We mark the position by drawing a small cross with a pencil.

8 Now we plot the next temperature, 8°C at 0900. We move along the x axis to the 0900 marker and go up until we're level with 8 on the y axis. Then we draw another cross.

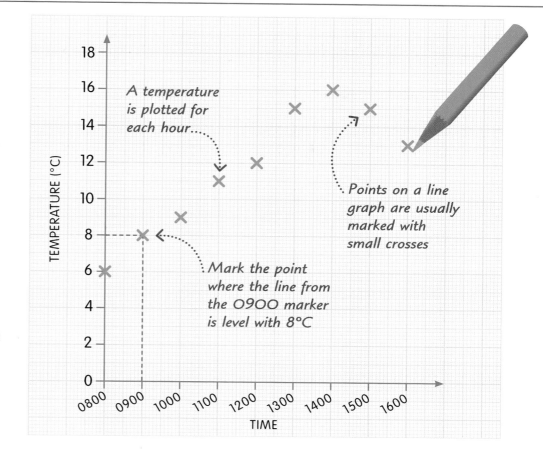

A temperature is plotted for each hour...

Points on a line graph are usually marked with small crosses

Mark the point where the line from the 0900 marker is level with 8°C

9 When we've plotted all the temperatures, we use a ruler to draw a straight line to link each pair of crosses. We do this between all the crosses on the graph, so that they're joined in an unbroken line.

10 Let's finish by giving our graph a title, so that anyone looking at it will know immediately what it's about.

Our line shows how the temperature rose during the morning and then started falling in the afternoon

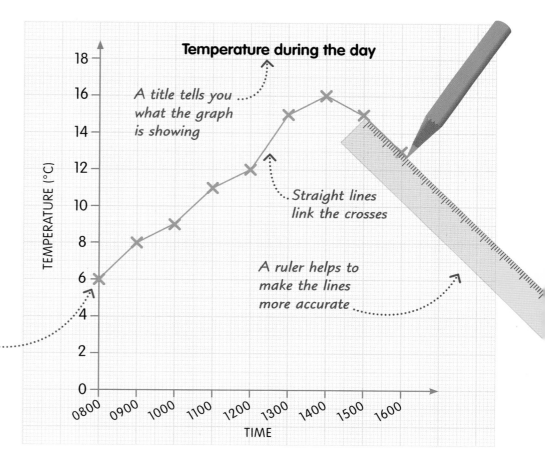

Temperature during the day

A title tells you what the graph is showing

Straight lines link the crosses

A ruler helps to make the lines more accurate

Pie charts

A pie chart presents information visually. It's a diagram that shows data as "slices", or sectors, of a circle. Pie charts are a good way of comparing the relative sizes of groups of data.

> The bigger the slice, the more data it represents.

1 Let's look at this pie chart. It shows the types of film that a group of school children said they most liked to watch.

2 Even though there are no numbers on the chart, we can still understand it. The bigger the sector, the more children chose that type of film.

3 We can compare the film types just by looking at the chart. It's clear that comedies are most popular and science fiction films are liked the least.

Favourite types of movie

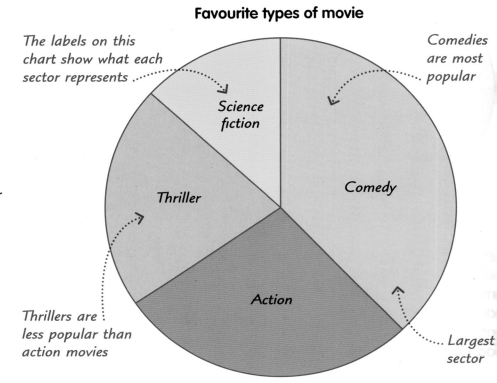

The labels on this chart show what each sector represents

Comedies are most popular

Comedies are most popular

Thrillers are less popular than action movies

Largest sector

Labelling sectors

There are two other ways of labelling pie charts: using a key or using labels.

KEY

● Science fiction

● Comedy

● Thriller

● Action

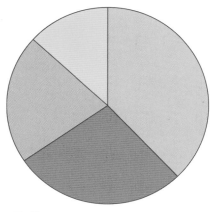

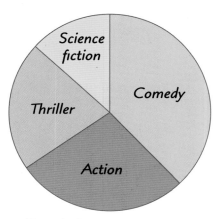

1 **Key**
We use the colours in the key to find out what type of film each sector represents.

2 **Labels**
We can also write our labels beside the chart or write them on the chart like here.

Pie-chart sectors

The circle, or "pie", is the whole set of data. Each of the sectors, or slices, is a subset. If we add up all the slices, we get the whole pie. We can express the size of a slice as an angle, a proper fraction, or a percentage.

1 Because it is a circle, a pie chart is a round angle of 360°. Each sector that makes up the chart takes up part of this bigger angle.

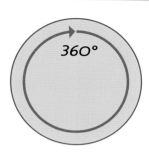

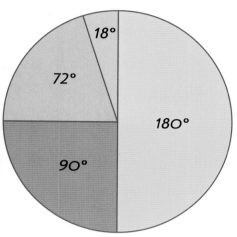

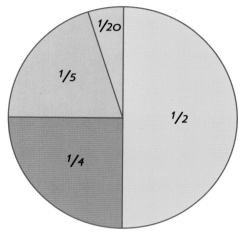

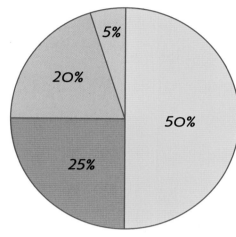

18 + 72 + 90 + 180 = 360°   ¹/20 + ¹/5 + ¹/4 + ¹/2 = 1   5% + 20% + 25% + 50% = 100%

2 Angles
The angle of a sector is measured from the centre in degrees (°). Together, the angles of the sectors always add up to 360°.

3 Fractions
Each sector is also a fraction of the chart. For example, a sector with an angle of 90° represents a quarter. Together, all the fractions add up to 1.

4 Percentages
Sectors may also be shown as percentages of the whole chart. A sector with an angle of 90° is 25 per cent. Together, the percentages add up to 100%.

TRY IT OUT

Pie-chart puzzles

Here are two problems to solve. Remember that the angles of a pie chart's sectors always add up to 360°, and when expressed as percentages the sectors always come to a total of 100%.

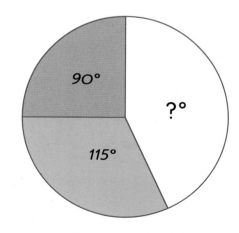

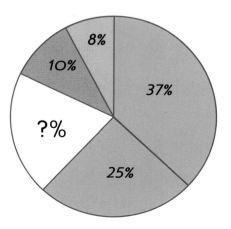

1 Can you work out the mystery angle of the third sector on this pie chart?

2 What's the percentage of the missing sector on this pie chart?

Answers on page 320

Making pie charts

We can make a pie chart from a frequency table of data using a pair of compasses and a protractor. There's a formula to help us to work out the angle of each sector, or "slice", on the chart.

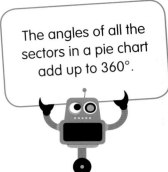

The angles of all the sectors in a pie chart add up to 360°.

Calculating the angles

The first step in drawing a pie chart is to calculate the angles of the slices.

1 Let's use the data in this frequency table to draw a pie chart. The sectors will represent the different flavours.

Ice cream sales	
Flavour	**Number sold**
Lemon	45
Mango	25
Strawberry	20
Mint	10
Total	**100**

Frequency (number of each flavour sold)

Total frequency (total number of ice creams sold)

2 To find the angles, we take the frequency for each flavour and put it into the formula on the right.

$$\text{Angle} = \frac{\text{frequency}}{\text{total frequency}} \times 360°$$

3 The table shows that out of 100 ice creams sold, 45 were lemon flavour. We can use these numbers in the formula to find the angle of the lemon sector: 45 ÷ 100 × 360 = 162°

Lemon ice creams sold (frequency)

Angle of lemon sector

$$\text{Lemon} = \frac{45}{100} \times 360° = 162°$$

Total number of ice creams sold (total frequency)

Angle of whole chart in degrees (°)

... 162°

4 Now we do the same for the other sectors. Then we add up all the angles to check that they come to 360°: 162 + 90 + 72 + 36 = 360°

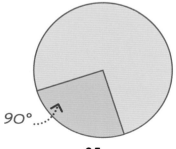

90°....

$$\text{Mango} = \frac{25}{100} \times 360° = 90°$$

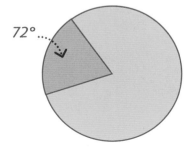

72° ....

$$\text{Strawberry} = \frac{20}{100} \times 360° = 72°$$

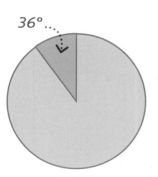

36°....

$$\text{Mint} = \frac{10}{100} \times 360° = 36°$$

Drawing the chart

Once we've found all the angles for the pie sectors, we're ready to make our chart. We'll need a protractor and a pair of compasses.

1 We'll draw a circle using a pair of compasses, so that it's accurate. We must make our circle big enough so that it's easy to colour in and put labels on.

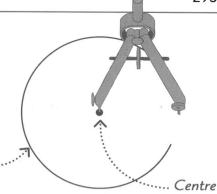

Draw the outline (circumference) of the circle

Centre

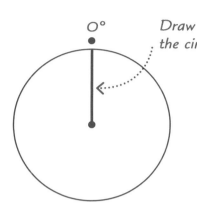

O°

Draw a line to the circle's edge

2 Let's draw a line from the centre to the circle's edge. We'll mark this as 0° and use it to measure our first angle.

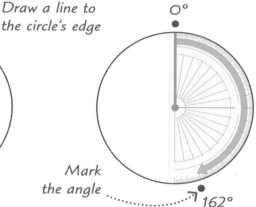

O°

Mark the angle 162°

3 Next, we put our protractor on our 0° line and use its scale to measure an angle of 162° for the lemon sector.

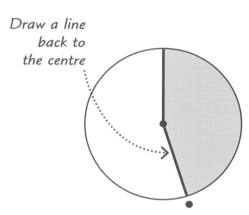

Draw a line back to the centre

4 Then we draw a line from the 162° angle back to the centre. The lemon sector is now complete. Let's colour it in.

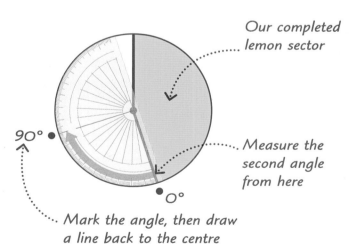

Our completed lemon sector

90°

Measure the second angle from here

O°

Mark the angle, then draw a line back to the centre

5 Now we align the protractor with the lower edge of the lemon sector and measure a 90° angle for the mango sector. We complete and colour in this sector.

Flavours of ice cream sold

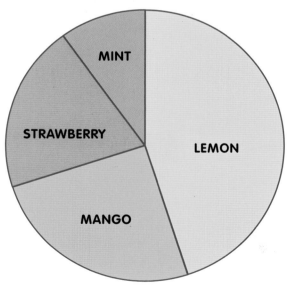

MINT

STRAWBERRY

LEMON

MANGO

6 We draw the remaining sectors in the same way. To finish off our chart, we add labels and a heading.

Probability

Probability is a measure of how likely something is to happen. It's often called chance. If something has a high probability, it's likely to happen. If something has a low probability, it's unlikely to happen. Probabilities are usually written as fractions.

Probability is the likelihood of something happening.

1 Let's think about tossing a coin. There are only two possible results: it will either land heads-up or tails-up.

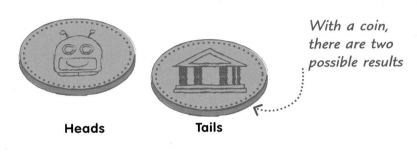

Heads   **Tails**

With a coin, there are two possible results

2 So what's the probability of tossing heads? Since you're just as likely to get heads as tails, there's an equal, or "even", chance of getting heads.

3 When you roll a dice, there are six possible results. So the probability of rolling a particular number, such as 3, is lower than getting heads in a coin toss.

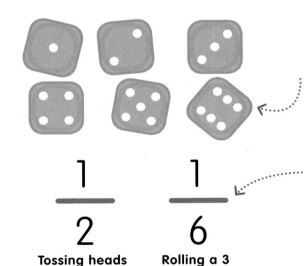

With dice, there are six possible results

A smaller fraction means a lower probability

4 We usually write probabilities as fractions. We say there's a 1 in 2 chance of getting heads in a coin toss, so we write it as ½. We've a 1 in 6 chance of rolling a 3 on a dice, so we write it as ⅙.

$$\frac{1}{2}$$

Tossing heads

$$\frac{1}{6}$$

Rolling a 3

REAL WORLD MATHS

Should I take my raincoat?

When meteorologists (weather scientists) make their forecasts, they include probability in their calculations. To predict whether or not it will rain, they look at previous days with similar conditions, such as air pressure and temperature. They work out on how many of those days it rained, and then they calculate the chance of rain today.

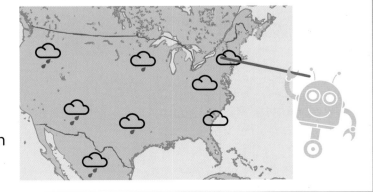

Probability scale

All probabilities can be shown on a line called a probability scale. The scale runs from 1 to 0. An event that's certain is 1, something that's impossible is 0. Everything else is in between these values.

1 We can be certain that the sun will rise tomorrow morning. Sunrise scores 1 and sits at the very top of the probability scale.

2 At this moment, it's very likely that somewhere around the world a plane is flying in the sky.

3 It's likely that at least one person among the pupils and staff at your school will have a birthday this week.

4 There is an equal chance of getting heads or tails when you toss a coin. Equal chance is the scale's halfway point.

5 It's unlikely that if you roll two dice you will throw a double six. As you'll know from board games, it doesn't happen often!

6 There's little chance of you being struck by a bolt of lightning. Although it's possible, it's very unlikely.

7 Flying elephants score 0 on the scale. Elephants don't have wings, so it's impossible to see a flying elephant.

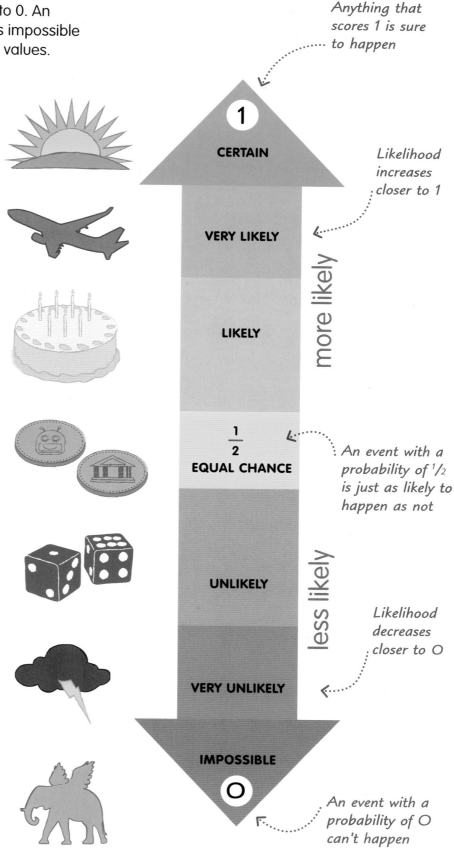

Anything that scores 1 is sure to happen

1
CERTAIN

Likelihood increases closer to 1

VERY LIKELY

LIKELY

more likely

$\frac{1}{2}$
EQUAL CHANCE

An event with a probability of $1/2$ is just as likely to happen as not

UNLIKELY

less likely

Likelihood decreases closer to 0

VERY UNLIKELY

IMPOSSIBLE

0

An event with a probability of 0 can't happen

Calculating probability

We can use a simple formula to help us work out the probability of something happening. The formula expresses the probability as a fraction. We can also change probability fractions into decimals and percentages.

1 Here's a box of 12 pieces of fruit. It contains six apples and six oranges, randomly arranged. What's the chance of picking out an apple if we shut our eyes?

2 Let's use the formula below to find the probability of choosing an apple:

number of results we're interested in
———————————————————————
number of all possible results

3 We can picture the formula like this. The top part of the formula means how many apples it's possible to take out of the box (6). The bottom part is the total number of fruits that could be chosen (12).

4 So, we've a 6 in 12 chance of picking an apple. We show this as the fraction 6/12, which can be simplified to 1/2.

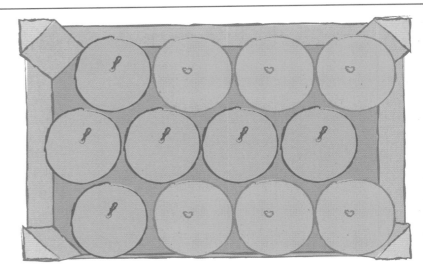

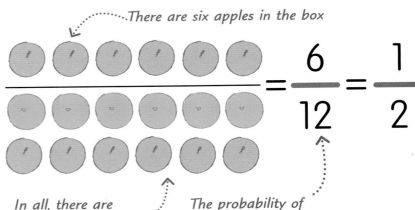

......There are six apples in the box

$$= \frac{6}{12} = \frac{1}{2}$$

In all, there are 12 fruits in the box......

The probability of picking an apple......

REAL WORLD MATHS

Unexpected results

Probability doesn't always tell us exactly what's going to happen. There's a 1 in 6 chance that this spinner will land on red. If we spin it 6 times, we'd expect to get a red at least once. But we might get 6 reds – or none.

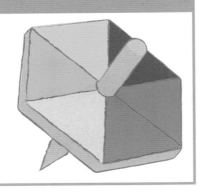

You can write probabilities as fractions, decimals, or percentages.

Decimals and percentages

Probabilities are most often written as fractions, but they can also be shown as decimals or percentages.

1 This box of 12 cakes contains three chocolate cakes and nine vanilla cakes. With our eyes shut, we have a 3 in 12 chance of choosing a chocolate cake.

2 Written as a fraction, the probability is ³/₁₂. We can simplify this to ¹/₄. Now we divide 1 by 4 to find the probability as a decimal: 1 ÷ 4 = 0.25. To change our decimal to a percentage, we simply multiply it by 100. So, 0.25 × 100 = 25%

Three chocolate cakes, nine vanilla cakes

Nine chocolate cakes, three vanilla cakes

3 Let's see what happens if the box contains nine chocolate cakes and three vanilla cakes.

4 Now the probability of picking a chocolate cake is ⁹/₁₂, or ³/₄. This is the same as 0.75 or 75%.

TRY IT OUT

Probability dice

Throwing dice is a great way to investigate probability. Dice throws are often important in board games, so if you know the probability of certain combinations occurring, you might be able to improve your gameplay!

Answers on page 320

1 What is the most likely total to occur when you roll two dice together? Start by writing down all the possible scores, and adding the numbers together.

2 What are the two least likely totals to occur?

3 What are the probabilities of getting the most likely and the least likely totals?

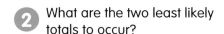

ALGEBRA

In algebra, we replace numbers with letters or other symbols. This makes it easier to study numbers and the connections between them – for example, to look at how they form patterns such as number sequences. By using algebra, we can also write helpful rules, called formulas, in a way that makes it easier to solve maths problems.

Equations

An equation is a mathematical statement that contains an equals sign. We can write equations using numbers, or with letters or other symbols to represent numbers. This type of maths is called algebra.

Balancing equations

An equation must always balance – whatever is to the left of the equals sign has the same value as whatever is to the right of the equals sign. We can see how this works when we look at this addition equation.

Both sides of this equation balance – they are equal

The three laws of arithmetic

An equation must always follow the three laws of arithmetic. We looked at how these rules work with real numbers on pages 154-55. We can also write the same laws using algebra if we replace the numbers with letters.

1 The commutative law
This law tells us that numbers can be added or multiplied in any order and the answer will always be the same. We can see how the commutative law works with this addition calculation, and then write the law using algebra.

The rules of arithmetic help to make sure that an equation balances.

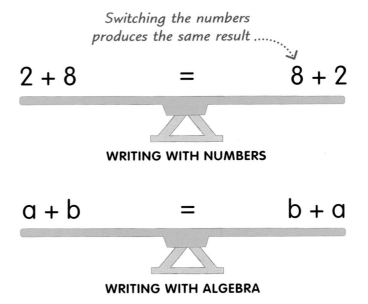

Switching the numbers produces the same result

$$2 + 8 \qquad = \qquad 8 + 2$$

WRITING WITH NUMBERS

$$a + b \qquad = \qquad b + a$$

WRITING WITH ALGEBRA

WRITING EQUATIONS WITH ALGEBRA

In algebra, we use some special words and phrases. We also write equations slightly differently compared with when we're using numbers.

In algebra, a number that we do not know yet can be represented by a letter. This is called a **variable**.	b
Instead of writing a × b, we simply write ab. We leave out the multiplication sign because it looks too much like the letter x.	ab
When we multiply numbers and letters, we write the number first.	4ab
A number, a letter, or a combination of both is called a **term**.	2b
Two or more terms separated by a maths sign is called an **expression**.	4 + c

2 The associative law

Remember, brackets tell us which part of a calculation to do first. This law tells us that when we're adding or multiplying, it doesn't matter where we put the brackets – the answer won't change. Take a look at this addition calculation.

Add the numbers within the brackets, then add 6 to get 13

$$(3 + 4) + 6 \quad = \quad 3 + (4 + 6)$$

WRITING WITH NUMBERS

$$(a + b) + c \quad = \quad a + (b + c)$$

WRITING WITH ALGEBRA

3 The distributive law

This is a law about multiplication. It says that adding a group of numbers together and then multiplying them by another number is the same as doing each multiplication separately and then adding them. Here's an example of how this law works.

Add the numbers within the brackets, then multiply the answer by 5

Multiply the numbers within the brackets, then add the answers

$$5 \times (2 + 4) = (5 \times 2) + (5 \times 4)$$

WRITING WITH NUMBERS

$$a (b + c) = ab + ac$$

WRITING WITH ALGEBRA

Solving equations

An equation can be rearranged to find the value of an unknown number, or variable.

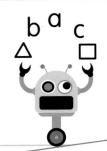

> It doesn't matter whether a shape or a letter represents the variable.

Simple equations

In algebra, a letter or a symbol represents the variable. We already know that the two sides of an equation must always balance. So, if the variable is on its own on one side of the equals sign, we can find its value by simply carrying out the calculation on the other side.

The shape represents the unknown value

1 Equations with symbols
Here we have two equations with a shape representing the unknown values. To find the answers we simply multiply or divide.

$$\triangle = 12 \times 7$$
$$\triangle = 84$$

$$\square = 72 \div 9$$
$$\square = 8$$

The letter represents the unknown value

2 Equations with letters
In these examples, letters are used to represent the unknown values. The equations are solved in the same way. We just follow the maths sign.

$$a = 36 + 15$$
$$a = 51$$

$$b = 21 - 13$$
$$b = 8$$

REAL WORLD MATHS

Everyday algebra

We use algebra every day without realizing it. For example, if we want to buy three bottles of juice, two boxes of cereal, and six apples, we can calculate the amount using an algebraic equation as shown here.

a = £2

b = £1

c = 50p

1 We write the equation as:
3a + 2b + 6c = total cost.

2 Now replace the letters with the prices as follows:
(3 × £2) + (2 × £1) + (6 × 50p) = £11

Rearranging equations

Finding the value of a variable is harder if the variable is mixed with other terms on one side of an equation. When this happens, we need to rearrange the equation so that the variable is by itself on one side of the equals sign. The key to solving the equation is to make sure it always balances.

Whatever we do to one side of the equation, we must do the same on the other side.

1 Let's look at this equation. We can solve it in simple stages so that we can isolate the letter b and find its value.

.... Variable

$$b + 25 = 46$$

2 Start by subtracting 25 from both sides and rewrite the equation. We know that 25 minus 25 equals zero. We say that the two 25s cancel each other out.

..... 25 and −25 cancel each other out

$$b + 25 - 25 = 46 - 25$$

3 We are left with the letter b on one side of the equals sign. We can now find its value by working out the calculation on the right of the equals sign.

.... The variable is now the subject of the equation

$$b = 46 - 25$$

4 When we work out 46 − 25, we are left with 21. So, the value of b is 21.

$$b = 21$$

5 We can check our answer by substituting 21 for the letter in the original equation.

.... Both sides of the equation balance

$$21 + 25 = 46$$

TRY IT OUT

Missing values

Can you simplify these equations to find the missing values?

1 $73 + b = 105$ **3** $i - 34 = 19$

2 $42 = 6 \times \square$ **4** $7 = \triangle \div 3$

Answers on page 320

Formulas and sequences

A sequence is a list of numbers that follows a pattern (see pages 14-17). By using a formula to write a rule for a sequence, we can work out the value of any term in the sequence without having to write out the whole list.

Number patterns

A number sequence follows a particular pattern, or rule. Each number in a sequence is called a term. The first number in a sequence is called the first term, the second number is called the second term, and so on.

In this sequence, each term is 2 more than the previous term

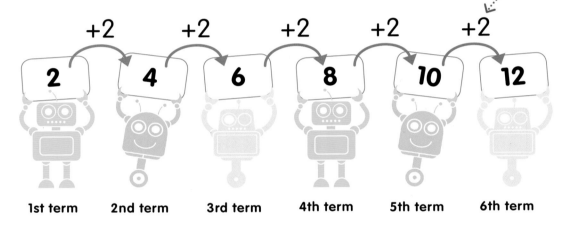

1st term   2nd term   3rd term   4th term   5th term   6th term

The nth term

In algebra, the value of an unknown term in a sequence is known as the nth term – the "n" stands for the unknown value. We can write a formula called a general term of the sequence to work out the value of any term.

The unknown term is called the nth term

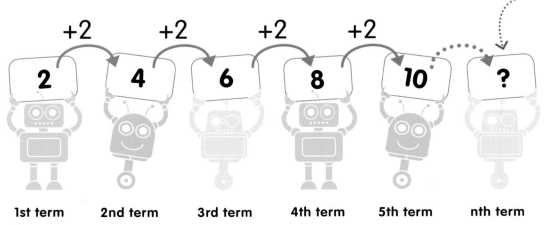

1st term   2nd term   3rd term   4th term   5th term   nth term

The dots show that the sequen goes on forever

Simple sequences

To find the formula for any sequence, we need to look at the pattern. Some sequences have an obvious pattern, so we can easily find the rule and write it as a formula.

The rule is multiply the term by 4

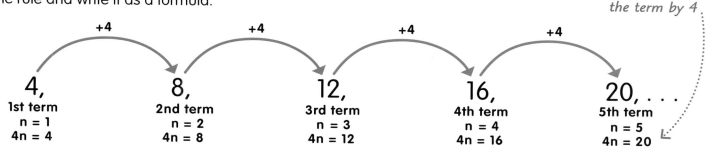

4,	**8,**	**12,**	**16,**	**20,** ...
1st term	2nd term	3rd term	4th term	5th term
n = 1	n = 2	n = 3	n = 4	n = 5
4n = 4	4n = 8	4n = 12	4n = 16	4n = 20

1 This sequence is made up of the multiples of 4. So, we can say the nth term is 4 × n. In algebra, we write this as 4n.

2 So, to find the value of the 30th term for example, we simply replace n in the formula with 30 and perform the calculation 4 × 30 = 120.

Two-step formulas

Some sequences will follow two steps such as multiplying and subtracting, or multiplying and adding.

The rule is multiply the term by 5, then subtract 1

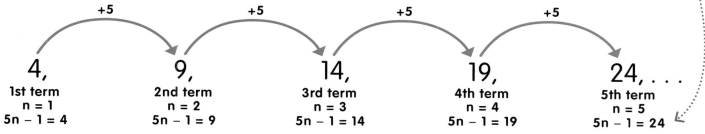

4,	**9,**	**14,**	**19,**	**24,** ...
1st term	2nd term	3rd term	4th term	5th term
n = 1	n = 2	n = 3	n = 4	n = 5
5n − 1 = 4	5n − 1 = 9	5n − 1 = 14	5n − 1 = 19	5n − 1 = 24

1 The formula for this sequence is 5n − 1. So, to find any term in the sequence, we have to perform a multiplication followed by a subtraction.

2 To find the 50th term in the sequence, for example, we replace n in the formula with 50. Then we can write 5 × 50 − 1 = 249. So, the 50th term is 249.

TRY IT OUT

Finding terms

The formula to work out the nth term in this sequence is 6n + 2. Can you continue the sequence and apply the formula?

Answers on page 320

8, 14, 20, 26, 32, 38,...

1 Write the next five numbers in this sequence.

2 Calculate the value of the 40th term.

3 Calculate the value of the 100th term.

Formulas

A formula is a rule for finding out the value of something. We write a formula using a combination of mathematical signs and letters to represent numbers or quantities.

In a formula, we can use letters instead of writing out all the words.

Writing a formula

A formula is like a recipe, except that in a formula we use signs and letters instead of words. A formula usually has three parts: a subject, an equals sign, and a combination of letters and numbers containing the recipe's instructions. Let's look at one of the simplest formulas, for finding the area of a rectangle. The formula is Area = length × width. Using algebra, we can write this as A = lw.

The subject of the formula

A

An equals sign shows the formula balances

=

The recipe (l × w)

lw

Using letters

Formulas use letters instead of words, so we need to know what the different letters stand for. Here are the letters we use to solve mathematical problems that involve measurement.

When we write a formula, we leave out the multiplication sign.

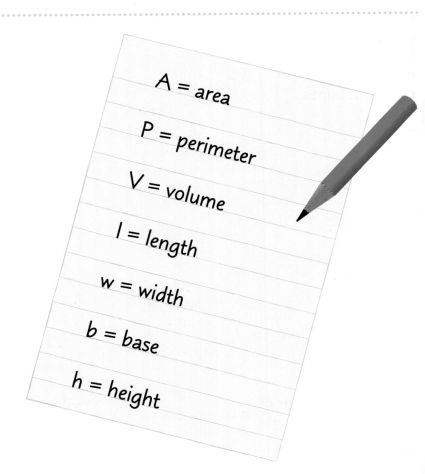

A = area

P = perimeter

V = volume

l = length

w = width

b = base

h = height

Using a formula

We use formulas in maths to find actual values. We can find the value of a formula's subject if we know the values of the variables on the other side of the equals sign.

WIDTH 3 m

The area is the space occupied by the swimming pool ....

LENGTH 5 m

1 We start by replacing the letters (A = lw) with the actual measurements. So, we have A = 5 × 3.

2 The length when multiplied by the width gives us 15. So, the area of this rectangular swimming pool is 15 m².

Common formulas

Here are some formulas you will need to know for finding the area, perimeter, and volume of some common shapes.

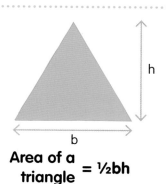

h

b

Area of a triangle = ½bh

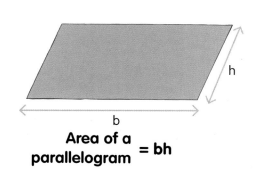

h

b

Area of a parallelogram = bh

The perimeter is the distance around the outside of a shape ....

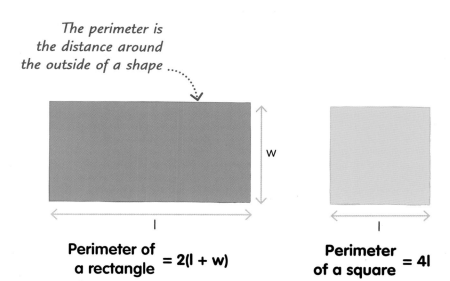

w

l

Perimeter of a rectangle = 2(l + w)

l

Perimeter of a square = 4l

The volume is the amount of space within a 3D shape ....

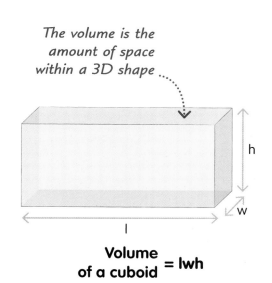

h

w

l

Volume of a cuboid = lwh

Glossary

acute angle An angle that is less than 90 degrees.

adjacent Next to each other, such as two angles or sides of a shape.

algebra The use of letters or other symbols to stand for unknown numbers when making calculations.

angle A measure of the amount of turn from one direction to another. You can also think of it as the difference in direction between two lines meeting at a point. Angles are measured in degrees. See *degree*.

anticlockwise Going round in the opposite direction to a clock's hands.

apex The tip or pointed top of any shape.

arc A curved line that forms a part of the circumference of a circle.

area The amount of space inside any 2D shape. Area is measured in square units, such as square metres.

associative law A law saying that if you add, for example, 1 + 2 + 3, it doesn't matter whether you add the 1 + 2 first or the 2 + 3 first. The law works for addition and multiplication, but not subtraction or division.

asymmetrical A shape with no reflective or rotational symmetry is asymmetrical.

average The typical or middle value of a set of data. There are different kinds of averages – see *mean*, *median*, and *mode*.

axis (plural *axes*) **(1)** One of the two main lines on a grid, used to measure the position of points, lines, and shapes. See also *x axis*, *y axis*. **(2)** An axis of symmetry is another name for a line of symmetry.

bar chart A diagram showing data as rectangular bars of different lengths or heights.

base The bottom edge of a shape, if you imagine it sitting on a surface.

block graph A diagram that shows data as stacks of square blocks.

brackets Symbols such as ( ) and [ ], used to surround numbers. They help show you which calculations you should do first.

capacity The amount of space inside a container.

Carroll diagram A diagram that is used to sort data into different boxes.

Celsius scale A scale of temperature. Water boils at 100 degrees on this scale.

centigrade scale Another name for the Celsius scale.

chord A straight line that cuts across a circle but doesn't go through the centre.

circumference The distance all the way round the outside of a circle.

clockwise Going round in the same direction as a clock's hands.

common denominator A term used when two or more fractions have the same lower number. See *denominator*.

common factor A factor that two or more numbers share. See *factor*.

common multiple A number that is a multiple of two or more different numbers. For example, 24 is a multiple of 3 as well as of 4, and so is a common multiple of these numbers. See *multiple*.

commutative law A law that says that, for example, 1 + 2 is the same as 2 + 1, and the order the numbers are in doesn't matter. It works for addition and multiplication, but not subtraction or division.

compass (1) An instrument that shows the direction of north, as well as other directions. **(2)** A pair of compasses is an instrument used to draw circles and parts of circles.

cone A 3D shape with a circular base and a side that narrows upwards to its apex. See *apex*

congruent Geometrical shapes that have the same size and shape.

conversion factor A number you multiply or divide by to change a measurement from one kind of unit to another. For example, if you've measured a length in metres and need to know it in feet, you have to multiply by 3.3.

coordinates Pairs of numbers that describe the position of a point, line, or shape on a grid or the position of something on a map.

cross section A new face made by cutting a shape parallel to one of its ends. See *face*.

cube number When you multiply a number by itself, and then by itself again, the result is called a cube number.

cubic unit Any unit, such as a cubic centimetre, for measuring the volume of a 3D shape. See *unit*.

cuboid A box-like shape with six faces, where opposite faces are identical rectangles.

cylinder A 3D shape with two identical circular ends joined by one curved surface. A tin can is an example.

data Any information that has been collected and can be compared.

decimal Relating to the number 10 (and to tenths, hundredths, and so on). A decimal fraction (also called a decimal) is written using a dot called a decimal point. The numbers to the right of the dot are tenths, hundredths, and so on. For example, a quarter (1/4) as a decimal is 0.25, which means 0 ones, 2 tenths, and 5 hundredths.

degree (symbol °) **(1)** A measure of the size of a turn or angle. A full turn is 360 degrees. **(2)** A unit on a temperature scale.

denominator The lower number in a fraction, such as the 4 in ³⁄₄.

diagonal (1) A straight line that isn't vertical or horizontal. **(2)** Inside a shape, a diagonal is any line joining two corners, or vertices, that aren't adjacent.

diameter A straight line from one side of a circle or sphere to the other that goes through the centre.

digit A single number from 0 to 9. Digits also make up larger numbers. For example, 58 is made up of the digits 5 and 8.

distributive law The law that says, for example, 2 × (3 + 4) is the same as (2 × 3) + (2 × 4).

dividend The number to be divided in a division calculation.

divisor The number you are dividing by in a division calculation.

equation A statement in maths that something equals something else, for example 2 + 2 = 4

equilateral triangle A triangle with all three sides and all three angles the same.

equivalent fraction A fraction that is the same as another fraction though it's written in a different way. For example, ²⁄₄ is equal to ½.

estimating Finding an answer that's close to the correct answer, often by rounding one or more numbers up or down.

face Any flat surface of a 3D shape.

factor A whole number that divides exactly into another number. For example, 4 and 6 are factors of 12.

factor pair Any two numbers that make a larger number when multiplied together.

Fahrenheit scale A scale of temperature. Water boils at 212 degrees on this scale.

formula A rule or statement that is written using mathematical symbols.

fraction A number that is not a whole number, for example ½, ¼, or ¹⁰⁄₃.

frequency (1) How often something happens. **(2)** In statistics, how many individuals or things have a particular feature in common.

gram (g) A unit of mass, a thousandth of a kilogram.

greatest common factor Another name for highest common factor.

grid method A way of multiplying using a grid drawn on paper.

highest common factor (HCF) The highest factor that two or more numbers have in common. For example, 8 is the highest common factor of 24 and 32.

horizontal Level and going from one side to the other, rather than up and down.

image A shape that's the mirror-image reflection of another shape, called the pre-image.

imperial units Traditional measuring units such as the foot, mile, gallon, and ounce. In science and maths, they have been replaced by metric units, which are easier to calculate with.

improper fraction A fraction that is greater than 1, for example ⁵⁄₂, which can also be written as the mixed number 2½. See *mixed number*.

intersect To meet or cross over (used of lines and shapes).

isosceles triangle A triangle with two sides the same length and two angles the same size.

kilogram (kg) The main unit of mass in the metric system, equal to 1000 grams.

kilometre (km) A metric unit of length, equal to 1000 metres.

lattice method A method of multiplying using a grid with diagonal lines on it.

line graph A diagram that shows data as points joined by straight lines. It's good for showing how measurements such as temperature can change over time.

line of reflection Also called the mirror line, a line exactly midway between an object and its reflection.

line of symmetry An imaginary line through a 2D shape that divides it into two identical halves. Some shapes have no line of symmetry, while others have several.

litre (l) A metric unit for measuring capacity.

long division A way of dividing by larger numbers that involves doing the calculation in stages.

long multiplication A written method for multiplying numbers with two or more digits. It involves doing the calculation in stages.

lowest common denominator The lowest common multiple of the denominators of different fractions. See *denominator*.

lowest common multiple The lowest number that is a common multiple of other given numbers. For example, 24 is a common multiple of 2, 4, and 6, but 12 is their lowest common multiple. See *multiple* and *common multiple*.

mass The amount of matter in an object. See *weight*.

mean An average found by adding up the values in a set of data and dividing by the number of values.

median The middle value of a set of data, when the values are put in order from lowest to highest.

metre (m) The main unit of length in the metric system, equal to 100 centimetres.

metric system A system of standard measuring units including the metre (for measuring length) and the kilogram (for measuring mass). Different measurements can be compared easily using these units by multiplying or dividing by 10, 100, or 1000.

milligram (mg) A metric unit of mass that equals a thousandth of a gram.

millilitre (ml) A metric unit of capacity that equals a thousandth of a litre.

millimetre (mm) A metric unit of length that equals one-thousandth of a metre.

mixed number A number that is partly a whole number and partly a fraction, such as $2\frac{1}{2}$.

mode The value that occurs most often in a set of data.

multiple Any number that's the result of multiplying two whole numbers together.

negative number A number less than zero: for example −1, −2, −3, and so on.

net A flat shape that can be folded up to make a particular 3D shape.

non-unit fraction A fraction with a numerator greater than one, for example $\frac{3}{4}$.

number A value used for counting and calculating. Numbers can be positive or negative, and include whole numbers and fractions. See *negative number, positive number*.

number line A horizontal line with numbers written on it, used for counting and calculating. Lowest numbers are on the left, highest ones on the right.

numeral One of the ten symbols from 0 to 9 that are used to make up all numbers. Roman numerals are different, and use capital letters such as I, V, and X.

numerator The upper number in a fraction, such as the 3 in $\frac{3}{4}$.

obtuse angle An angle between 90 and 180 degrees.

operator A symbol that represents something you do to numbers, for example + (add) or × (multiply).

opposite angles The angles on opposite sides where two lines intersect, or cross over each other. Opposite angles are equal.

origin The point where the x and y axes of a grid intersect.

parallel Running side by side without getting closer or further apart.

parallelogram A type of quadrilateral whose opposite sides are parallel and equal to each other.

partitioning Breaking numbers down into others that are easier to work with. For example, 36 can be partitioned into 30 + 6.

percentage (%) A proportion expressed as a fraction of 100 – for example, 25 per cent (25%) is the same as $\frac{25}{100}$.

perimeter The distance around the edge of a shape.

perpendicular Something is perpendicular when it is at right angles to something else.

pictogram A diagram that shows data as rows or columns of small pictures.

pie chart A diagram that shows data as "slices" (sectors) of a circle.

place-value system Our standard way of writing numbers, where the value of each digit in the number depends on its position within that number. For example, the 2 in 120 has a place value of twenty, but in 210 it stands for two hundred.

polygon Any 2D shape with three or more straight sides, such as a triangle or a parallelogram.

polyhedron Any 3D shape whose faces are polygons.

positive number A number greater than zero.

prime factor A factor that is also a prime number. See *factor*.

prime number A whole number greater than 1 that can't be divided by any whole number except itself and 1.

prism A 3D shape whose ends are two identical polygons. It is the same size and shape all along its length.

probability The chance of something happening or being true.

product The number you get when you multiply other numbers together.

proper fraction A fraction whose value is less than 1, where the numerator is less than the denominator, for example $\frac{2}{3}$.

proportion The relative size of part of something, compared with the whole.

protractor A tool, usually made of flat, see-through plastic, for measuring and drawing angles.

quadrant A quarter of a grid when the grid is divided by x and y axes.

quadrilateral A 2D shape with four straight sides.

quotient The answer you get when you divide one number by another.

radius Any straight line from the centre of a circle to its circumference.

range The spread of values in a set of data, from the lowest to the highest.

ratio Ratio compares one number or amount with another. It's written as two numbers, separated by a colon (:).

rectangle A four-sided 2D shape where opposite sides are the same length and all the angles are 90 degrees.

reflection A type of transformation that produces a mirror image of the original object. See *transformation*.

reflective symmetry A shape has reflective symmetry if you can draw a line through it to make two halves that are mirror images of each other.

reflex angle An angle between 180 and 360 degrees.

remainder The number that is left over when one number doesn't divide into another exactly.

rhombus A quadrilateral with all four sides the same length. A rhombus is a special kind of parallelogram, in which all the sides are of equal length. See *parallelogram.*

right angle An angle of 90 degrees (a quarter turn), such as the angle between vertical and horizontal lines.

right-angled triangle A triangle where one of the angles is a right angle.

rotation Turning around a central point or line.

rotational symmetry A shape has rotational symmetry if it can be turned around a point until it fits exactly into its original outline.

rounding Changing a number to a number, such as a multiple of 10 or 100, that's close to it in value and makes it easier to work with.

scalene triangle A triangle where none of the sides or angles are the same size.

sector A slice of a circle similar in shape to a slice of cake. Its edges are made up of two radii and an arc.

segment (1) Part of a line. **(2)** In a circle, the area between a chord and the circumference.

sequence An arrangement of numbers one after the other that follows a set pattern, called a rule.

set A collection or group of things, such as words, numbers, or objects.

significant digits The digits of a number that affect its value the most.

simplify (a fraction) To put a fraction into its simplest form. For example, you can simplify $^{14}/_{21}$ to $^2/_3$.

solid In geometry, a term for any 3D shape, including a hollow one.

sphere A round, ball-shaped 3D shape, where every point on its surface is the same distance from the centre.

square A four-sided 2D shape where all the sides are the same length and all the angles are 90 degrees. A square is a special kind of rectangle. See *rectangle.*

square number If you multiply a number by itself, the result is called a square number, for example $4 \times 4 = 16$

square unit Any unit for measuring the size of a flat area. See *unit.*

straight angle An angle of exactly 180 degrees.

subset A set that is part of a larger set. See *set.*

symmetry A shape or object has symmetry if it looks exactly the same after a reflection or rotation.

tally marks Lines drawn to help record how many things you've counted.

tangent A straight line that just touches a curve or the circumference of a circle at a single point.

three-dimensional (3D) Having length, width, and depth. All solid objects are three-dimensional – even very thin paper.

ton/tonne A tonne is a metric unit of mass equal to a thousand kilograms: it is also called a metric ton. A ton is also a traditional imperial unit, which is almost the same size as a tonne.

transformation Changing the size or position of a shape or object by reflection, rotation, or translation.

translation Changing the position of a shape or object without rotating it or changing its size or shape.

trapezium A quadrilateral with one pair of sides parallel, also called a trapezoid.

triangle A 2D shape with three straight sides and three angles.

turn To move round a fixed point, such as hands moving on a clock.

two-dimensional (2D) Having length and width, or length and height, but no thickness.

unit A standard size used for measuring, such as the metre (for length) or the gram (for mass).

unit fraction A fraction in which the numerator is 1, for example $^1/_3$.

universal set The set that includes all the data you're investigating. See *set.*

value The amount or size of something.

variable An unknown number in an equation. In algebra, a variable is usually represented by a letter or a shape.

Venn diagram A diagram that shows sets of data as overlapping circles. The overlaps show what the sets have in common.

vertex (plural *vertices*) An angled corner of a 2D or 3D shape.

vertical Going in a straight up and down direction.

volume The three-dimensional size of an object.

weight A measurement of the force of gravity acting on an object. See *mass.*

whole number Any number such as 8, 36 or, 5971 that is not a fraction.

x axis The horizontal line that is used to measure the position of points plotted on a grid or graph.

y axis The vertical line that is used to measure the position of points on a grid or graph.

Index

Answers

Numbers

p11 **1)** 1998 **2)** MDCLXVI and MMXV

p15 **1)** 67, 76 **2)** 24, 28 **3)** 92, 90 **4)** 15, 0

p19 **1)** 10 **2)** −5 **3)** −2 **4)** 5

p21 **1)** 5123 < 10 221
2) −2 < 3
3) 71 399 > 71 000
4) 20 − 5 = 11 + 4

p23 Trevor 1, Bella 3, Buster 7, Jake 9, Anna 13, Uncle Dan 35, Mum 37, Dad 40, Grandpa 67, Grandma 68

p27 **1)** 170 cm **2)** 200 cm

p31 **multiples of 8:** 16, 32, 48, 56, 64, 72, 144
multiples of 9: 18, 27, 36, 72, 81, 90, 108, 144
common multiples: 72, 144

p35 Here is one of the ways to complete the factor tree:

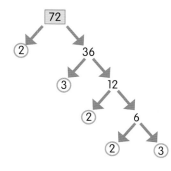

p38 **1)** 100 **2)** 16 **3)** 9

p47 18 chickens.

p51 Wook got the most right: he got $^{25}/_{30}$ correct to Zeek's $^{24}/_{30}$

p57 **1)** $^{1}/_{12}$ **2)** $^{1}/_{10}$ **3)** $^{1}/_{21}$ **4)** $^{1}/_{6}$

p61 Twerg 17.24, Bloop 16.56, Glook 17.21, Kwonk 16.13, Zarg 16.01.
Zarg's time is fastest.

p63 **1)** 4.1 **2)** 24.4 **3)** 31.8 **4)** 20.9

p65 **1)** 25% **2)** 75% **3)** 90%

p66 **1)** 60% **2)** 50% **3)** 40%

p67 **1)** 20 **2)** 55 **3)** 80

p69 **1)** £100 **2)** £35 **3)** £13.50

p73 The T. rex is 560 cm (5.6 m) high and 1200 cm (12 m) long.

p75 **1)** $^{35}/_{100}$ simplified to $^{7}/_{20}$ **2)** 3%, 0.03 **3)** $^{4}/_{6}$ simplified to $^{2}/_{3}$

Calculating

p82 **1)** 100 **2)** 1400 **3)** 100 **4)** 1 **5)** 100 **6)** 10 000

p85 **1)** 823 **2)** 1590 **3)** 11 971

p87 **1)** 8156 **2)** 9194 **3)** 71.84

p90 **1)** 800 **2)** 60 **3)** 70 **4)** 70 **5)** 0.02 **6)** 0.2

p91 377

p93 **1)** £6.76 **2)** £2.88 **3)** £40.02

p95 **1)** 207 **2)** 423 **3)** 3593

p99 **1)** 24 **2)** 56 **3)** 54 **4)** 65

p101 **1)** 1,14 ; 2,7
2) 1,60 ; 2,30 ; 3,20 ; 4,15 ; 5,12 ; 6,10
3) 1,18 ; 2,9 ; 3,6
4) 1,35 ; 5,7
5) 1,24 ; 2,12 ; 3,8 ; 4,6

p103 **1)** 28, 35, 42
2) 36, 45, 54
3) 44, 55, 66

p105 52, 65, 78, 91, 104, 117, 130, 143, 156

p108 **1)** 679 **2)** 480 000 **3)** 72

p109 **1)** 1250 **2)** 30 **3)** 6930 **4)** 3010 **5)** 2.7 **6)** 16 480

p111 **1)** 770 **2)** 238 **3)** 312 **4)** 1920

p115 3072

p117 **1)** 2360 **2)** 4085 **3)** 8217 **4)** 16 704 **5)** 62 487

p131 **1)** £9 each **2)** 6 marbles each

p133 **1)** 12 **2)** 8 **3)** 6 **4)** 4 **5)** 3 **6)** 2

p136 **1)** £182.54 **2)** 4557 cars

p137 **1)** 43 leaflets **2)** 45 bracelets

p141 **1)** 32 r4 **2)** 46 r4

p143 **1)** 31 **2)** 71 r2 **3)** 97 r2 **4)** 27 r4

p145 **1)** 151 **2)** 2

p153 **1)** 37 **2)** 17 **3)** 65

p157 **1)** 1511 **2)** 2.69 **3)** −32 **4)** 2496 **5)** 17 **6)** 240

Measurement

p162 50 m

p164 **1)** 87 cm **2)** 110 cm

p168 **1)** 16 cm² **2)** 8 cm² **3)** 8 cm²

p170 8 m²

p171 3 m

p175 77 m²

p180 **1)** 15 cm³ **2)** 20 cm³ **3)** 14 cm³

p181 1 000 000 (1 million)

p184 7 g

p185 13 360 g or 13.36 kg

p187 26°C

p197 70 minutes

p201 £9.70

Geometry

p207 There are nine diagonals:

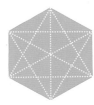

p209 The dotted lines show parallel lines:

p213 Shape 1 is the regular polygon.

p215

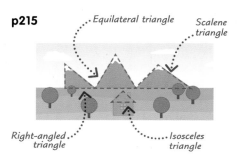

Equilateral triangle *Scalene triangle*

Right-angled triangle *Isosceles triangle*

p217 You would get a parallelogram.

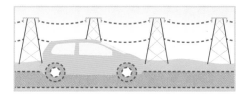

p221 The diameter is 6 cm. The circumference is 18.84 cm.

p223 The shape has 8 faces, 18 edges, and 12 vertices.

p227 Shape 4 is a non-prism.

p228 The other nets of a cube are:

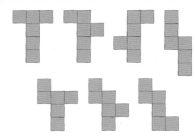

p237 $a = 90°$, $b = 50°$, c and $e = 40°$

p239 **1)** 30° **2)** 60°

p241 Each angle is 70°

p243 **1)** 60° **2)** 34° **3)** 38° **4)** 55°

p247 115°

p248 A = (1,3) B = (4,7)
C = (6,4) D = (8,6)

p251 **1)** (2, 0), (1, 3), (−3, 3), (−4, 0), (−3, −3), (1, −3).
2) You would make this shape:

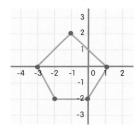

p253 **1)** Orange monorail car
2) Boat no. 2 **3)** C7

p255 **1)** 2W, 2N, 3W
2) One route is: 2E, 8N, 1E
3) The beach **4)** Seal Island

p257 The numbers 7 and 6 have none, 3 has one, and 8 has two.

p258 No. 3 has no rotational symmetry.

p261

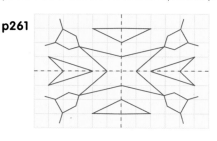

p265 There are five other positions the triangle could be in:

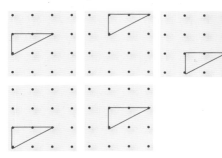

Statistics

p277 **1)** 133 **2)** 7 **3)** 19

p283 One of several possible pictograms looks like this:

Leroy's gaming	
Day	**Gaming time**
Monday	🎮🎮🎮
Tuesday	🎮🎮🎮🎮🎮🎮
Wednesday	🎮🎮
Thursday	🎮🎮🎮🎮🎮
Friday	🎮🎮🎮🎮🎮🎮🎮

KEY
🎮 10 minutes

p293 **1)** 155° **2)** 20%

p299 **1)** 7 **2)** 2 and 12 **3)** $\frac{1}{6}$ and $\frac{1}{36}$

Algebra

p305 **1)** 32 **2)** 7 **3)** 53 **4)** 21

p307 **1)** 44, 50, 56, 62, 68 **2)** $(6 \times 40) + 2 = 242$ **3)** $(6 \times 100) + 2 = 602$

Acknowledgments

Dorling Kindersley would like to thank: Thomas Booth for editorial assistance; Angeles Gavira-Guerrero, Martyn Page, Lili Bryant, Andy Szudek, Rob Houston, Michael Duffy, Michelle Baxter, Clare Joyce, Alex Lloyd, and Paul Drislane for editorial and design work on early versions of this book; Kerstin Schlieker for editorial advice; and Iona Frances, Jack Whyte, and Hannah Woosnam-Savage for help with testing.